Ultracool Dwarfs
New Spectral Types L and T

Springer
Berlin
Heidelberg
New York
Barcelona
Hong Kong
London
Milan
Paris
Tokyo

Physics and Astronomy ONLINE LIBRARY

http://www.springer.de/phys/

Hugh R. A. Jones   Iain A. Steele (Eds.)

Ultracool Dwarfs

New Spectral Types L and T

With 104 Figures and 11 Tables

Springer

Dr. Hugh R.A. Jones
Dr. Iain A. Steele
Astrophysics Research Institute
Twelve Quays House
Egerton Wharf
Birkenhead CH41 1LD
United Kingdom
E-mail: hraj@astro.livjm.ac.uk
       ias@astro.livjm.ac.uk

Library of Congress Cataloging-in-Publication Data applied for
Die deutsche Bibliothek - CIP-Einheitsaufnahme
Ultracool dwarfs : new spectral types L and T ; with 11 tables / Hugh R. A. Jones ; Iain A. Steele (ed.).
- Berlin ; Heidelberg ; New York ; Barcelona ; Hong Kong ; London ; Milan ; Paris ; Tokyo : Springer, 2001
(Physics and astronomy online library)
ISBN 3-540-42353-2

ISBN 3-540-42353-2 Springer-Verlag Berlin Heidelberg New York

This work is subject to copyright. All rights are reserved, whether the whole or part of the material is concerned, specifically the rights of translation, reprinting, reuse of illustrations, recitation, broadcasting, reproduction on microfilm or in any other way, and storage in data banks. Duplication of this publication or parts thereof is permitted only under the provisions of the German Copyright Law of September 9, 1965, in its current version, and permission for use must always be obtained from Springer-Verlag. Violations are liable for prosecution under the German Copyright Law.

Springer-Verlag Berlin Heidelberg New York
a member of BertelsmannSpringer Science+Business Media GmbH

http://www.springer.de

© Springer-Verlag Berlin Heidelberg 2001
Printed in Germany

The use of general descriptive names, registered names, trademarks, etc. in this publication does not imply, even in the absence of a specific statement, that such names are exempt from the relevant protective laws and regulations and therefore free for general use.

Typesetting: Camera-ready copy from the authors using a Springer TEX macro package
Cover design: *design & production* GmbH, Heidelberg

Printed on acid-free paper   SPIN 10844749   55/3141/di   5 4 3 2 1 0

Preface

Once you have looked at the night sky on a moonless night it is not hard to realise why so much of our science and religion has its roots in the stars. Yet it took until 1850 to realise that fainter stars were not necessarily further away, nor the brighter ones closer. In fact within the magnitude range observable to the naked eye it is probable that the brighter star is in fact further away. Even today the measurement of stellar distances is relatively difficult and is generally only done using dedicated telescopes. In the early years of the 20th century Hertzsprung and Russell developed a powerful classification diagram which allows stars to be distinguished using a plot of their colour versus magnitude. The construction of this diagram involved the use of spectroscopy which has become the cornerstone of modern astronomy. As telescopes become more powerful, detectors more sensitive and more physics is added to astrophysics, astronomical spectroscopy becomes a more powerful tool.

The concern of this book is the spectral classification of stars. With a single spectrum of a star it is possible to uniquely classify an object and find its place on the Hertzsprung–Russell diagram. This spectrum is thus equivalent to having the colour and the magnitude of the object which can in turn be related to mass and other quantities. More than this a stellar spectrum allows insight into the chemical composition and age of a star. Thus the spectral typing of stars using spectroscopy is central to astrophysics. Stars are traditionally classified into the main spectral types O, B, A, F, G, K, M, R and S, with each of the spectral types being broken down into a number of increments, usually 1 to 8, and furthermore being allocated a luminosity class I to VIII. This range of sub-divisions allows all known stars to be given a unique spectral type. This system has been tweaked over the last 100 years but has altered little since the 1930s. This book arises because since the mid-1990s a number of objects have been discovered which do not fit within the existing spectral classification scheme. Here we are interested in the properties of these objects and the new spectral types and classification systems necessary to understand them.

Ultracool dwarfs extend from the coolest M dwarfs into the brown dwarf regime and the new spectral types L and T. It is likely that most of the ultracool dwarfs are not massive enough to undergo nuclear fusion in their cores to burn hydrogen and are thus brown dwarfs rather than stars.

Like any form of classification, spectral typing can easily become bogged down in questions of standardization and nomenclature and there are indeed different methods presented within this book; however the proposed systems of spectral classification for ultracool dwarfs are trying to avoid this by as far as possible continuing the same philosophies and techniques used by previous generations of spectroscopists. This traditional approach is likely to prove particularly helpful in understanding the properties and the importance of the divide between stars and brown dwarfs. On the other hand it will be probably be necessary to develop new systems of spectral classification, particularly as we attempt to classify objects cooler than 1000 K.

The fact that after 70 years new spectral types are required precipitated the International Astronomical Union's Commissions 27 and 34 to call a meeting within the 34th General Assembly of the International Astronomical Union, during August 2000, to discuss the properties and rationale of ultracool dwarfs. This book has arisen from the talks given at this meeting and serves to record the state of this new field at the beginning of a new millennium. For convenience we have divided the manuscript into three parts ("Theory", "Observations" and "Spectral Classification"); however, as for all modern astrophysics, the distinctions are blurred, and the reader will certainly find important comments regarding each of these topics in each part.

Finally we wish to thank the staff and Commission members of the IAU for their help both before and during the meeting, and especially Prof. Michele Gerbaldi for her assistance in getting the meeting approved and organized.

Liverpool,
July 2001

Hugh Jones,
Iain Steele
Liverpool John Moores University

Contents

Part I   Theory

Introduction: Theoretical Models of Brown Dwarfs –
an Observer's Perspective
J. Liebert 3

Unified Model Photospheres for Ultracool Dwarfs
of Types L and T
T. Tsuji 9

Alkali Metals and the Colour of Brown Dwarfs
A. Burrows 26

Formation of the Optical Spectra of L Dwarfs
Y. Pavlenko 33

Part II   Observations

Introduction: The Coolest Dwarfs – a Brief History
R.F. Jameson 53

Imaging and Spectroscopy
of Hot (Young) "Ultracool" Companions
*G. Schneider, P.J. Lowrance, E.E. Becklin, J.D. Kirkpatrick, P. Plait,
S.R. Heap, E. Malumuth, R.J. Terille, C. Dumas, A.B. Schultz,
B.A. Smith, A.J. Weinberger, D.C. Hines* 56

Activity and Kinematics of M and L Dwarfs
J.E. Gizis 71

Infrared Spectroscopy of Brown Dwarfs:
the Onset of CH_4 Absorption in L Dwarfs and the L/T Transition
*T.R. Geballe, K.S. Noll, S.K. Leggett, G.R. Knapp, X. Fan,
and D. Golimowski* 83

**Surface Features, Rotation and Atmospheric Variability
of Ultra Cool Dwarfs**
C.A.L. Bailer-Jones 92

Low-Mass Stellar and Brown Dwarf Binary Systems
I.N. Reid, D.W. Koerner, J.E. Gizis, J.D. Kirkpatrick 111

The Second Guide Star Catalogue and Cool Stars
R.L. Smart, D. Carollo, M.G. Lattanzi, B. McLean, A. Spagna 119

**Low-Luminosity Companions to Nearby Stars:
Status of the 2MASS Data Search**
*J.D. Kirkpatrick, J.E. Gizis, A.J. Burgasser, J.C. Wilson, C.C. Dahn,
D.G. Monet, I.N Reid, J. Liebert* 125

Part III Spectral Classification

Introduction: The Spectral Types of the Ultracool Dwarfs
M.S. Bessell 135

The Classification of L Dwarfs
J.D. Kirkpatrick 139

**Spectroscopy of Young Brown Dwarfs
and Isolated Planetary Mass Objects**
E.L. Martín 153

The Classification of T Dwarfs
A.J. Burgasser, J.D. Kirkpatrick, M.E. Brown 169

**L-Band Photometry and Spectroscopy of L and T Dwarfs:
Exploring Infrared Spectral Typing**
D.C. Stephens, M.S. Marley, K.S. Noll 183

Index 195

Part I

Theory

Introduction: Theoretical Models of Brown Dwarfs – an Observer's Perspective

J. Liebert

Department of Astronomy and Steward Observatory,
The University of Arizona, Tucson, AZ 85721, USA

Abstract. The three papers presented in this section on model atmospheres of brown dwarfs raise several common themes. I discuss three of these from an observational viewpoint – the temperature scale for L dwarfs, the role of the alkali elements, and the likely "weather" in brown dwarf atmospheres.

1 Introduction

The results reported in this section by Tsuji, Burrows, and Pavlenko demonstrate the rapid progress theorists are making in addressing the many new observations of L and T dwarfs of the last few years. I thought it would be helpful just to give an observer's perspective of some main issues which may be emerging. I will discuss three of these.

2 The L Dwarf Temperature Scale

I begin by showing (Fig. 1) an Hertzsprung-Russell Diagram – M_J vs. $I-J$ colour – of K through T dwarfs with trigonometric parallaxes (or brighter companions with same). This figure is similar to Fig. 2 of Reid et al. [14]. The caption explains the symbols for two samples of K and M dwarfs. The impressive new parallaxes of L dwarfs (filled circles) all come from our colleagues at the U.S. Naval Observatory, Flagstaff Station, a program directed by Conard C. Dahn, with David G. Monet responsible (mostly) for the reduction procedures. USNO participates in both the 2MASS and SDSS projects, and the L dwarf targets come from these two sources and the DENIS project. The T dwarfs have been too faint and recently-discovered for any parallaxes up to now, but the two important companion T dwarfs, Gl 229B and Gl 570D, have excellent parallaxes from the bright primaries.

Neill Reid chose the I and J bands because this colour appears to change in a monotonic way with decreasing luminosity and (apparently) $T_{\rm eff}$. In contrast, the H and K bands are both strongly affected by the onset of CH_4 absorption and other opacities at the L/T transition. The monotonic relationship between the absolute (M_J) magnitude and colour for K through L dwarfs is indeed impressive. The M's end where the L's start and the latter end where the T's start. Kirkpatrick et al. [6,7] find that this correlation is strong between the L subtype and various absolute magnitudes and colours. That is, the L spectral typing does

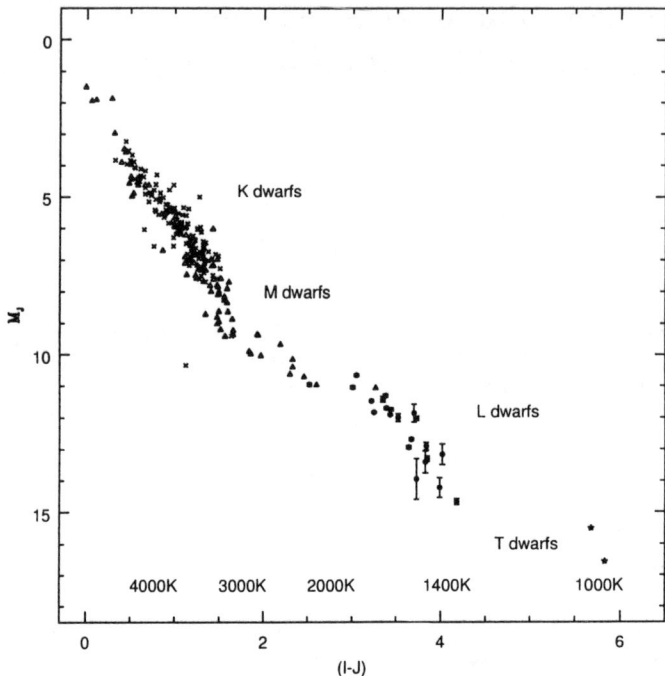

Fig. 1. The M_J, $I_C - J$ diagram for K-T dwarfs with direct distance measurements. Crosses are nearby stars with BVRI data from Bessell [3]), JHK observations from 2MASS and parallax measurements from Hipparcos; open triangles mark nearby, single stars with accurate trigonometric parallaxes (Reid and Gizis [13]), and late-M dwarfs from the USNO parallax program (Dahn et al., in prep.); filled circles are L dwarfs from this program. Rough estimates of $T_{\rm eff}$ across the bottom mark the spectral type boundaries.

a good job of rank ordering these objects by $T_{\rm eff}$. This monotonic nature of the L dwarf ordering is emphasized in Tsuji's paper. Now the big task is determining what the upper and lower bounds, and range of $T_{\rm eff}$ actually are.

Kirkpatrick et al. [6,7] have used the appearance/disappearance of individual spectral features such as TiO, VO and CH_4 as the basis for a scale running from 2000/2100 K (type L0) to 1300/1400 K for L8 dwarfs (the scale indicated at the bottom of Fig. 1). As predcited by Tsuji et al. [17], the top of the range is set by the weakening of TiO and VO, and is the starting point of not only his discussion here, but also the papers of Pavlenko and Burrows. As all of us now know, the appearance of CH_4 predicted near the bottom of this range signals the end of L dwarfs and the L/T transition. In contrast, Basri et al. [2] determine temperatures from fitting high resolution alkali line profiles with model atmosphere models. They derive a hotter scale of 2200 K to 1700 K. In favoring a 700 K "gap" between the coolest known L dwarf and Gl 229B, they argue that later, yet-to-be- discovered L subtypes and many L/T transition objects must occupy this region of temperature. It is therefore interesting that Dr. Pavlenko's attempts to fit DENIS J0205-1159, one of the latest L dwarfs, favor a T_{eff} of 1200-1400 K (see also [11]).

Indeed, there may be good reasons to believe that the actual gap between L8 and classical T dwarfs is much smaller than 700 K. Returning to Fig. 1, I point out the mere half-magnitude difference between the latest L's and Gl 229B in the M_J magnitude. For reasons discussed previously, one might therefore expect M_J to provide a reliable measure of M_{bol}. This is equivalent to making the assumption that the bolometric correction (BC_J) between M_J and M_{bol} (the total luminosity) is varying slowly between late L and T. Indeed, multiwavelength observations of Gl 229B show that it has BC_J of 2.2 magnitudes, and hence M_{bol} = 17.7. Yet even late M dwarfs have BC_J values only 0.2 mags different. Hence, the close proximity of late L and T dwarfs in M_J is evidence that the temperature gap is also small. It is likely that a few L/T transition objects populate this temperature region. For a more detailed presentation of these arguments, please see Reid [12].

3 The Role of the Alkali Metals

All three of the theorists discuss at length the dominant role of the alkali resonance doublets and subordinate lines in the optical and near-infrared spectra of L and T dwarfs. No longer need one invoke a haze, or wavelength-dependent dust absorption, to explain the observed slopes of T dwarf energy distributions shortward of 1 μm. Pavlenko's discussion of the broadening parameters for these lines is a little bit frightening and harkens back to the earlier paper of Burrows, Marley and Sharp [4]. When lines become thousands of Angstroms wide, the physics of the broadening becomes complicated! One wonders whether to believe T_{eff} estimates from fitting these lines. Fortunately, the Basri technique relies on much weaker lines like Cs, rather than those of K or Na.

I cannot resist inserting into these proceedings the best currently available "optical" spectrum (down to 0.6 μm) of any T dwarf (Fig. 2). This is our Keck II LRIS [10] spectrum of SDSS 1624+0029 (Liebert et al. [9]), Strauss et al.'s first field T dwarf [15]. It is wonderful to predict that the slope of the red spectrum is dominated by the red wing of K I stretching thousands of Angstroms, but it is also nice to prove this by observing directly the strong line core. The downturn below 7000Å is likely evidence for an even-stronger Na I doublet, centered unfortunately just bluewards of the edge of the spectrum. I invite all of our theorists to fit this spectrum – the only such spectrum so far for a T dwarf – and the infrared spectra. Our USNO colleagues should announce a trigonometric parallax shortly for this object.

4 Brown Dwarf Weather

Earlier I have emphasized the monotonic behavior of the L dwarfs, that the typing rank-orders with colour and absolute magnitude. It remains to be seen

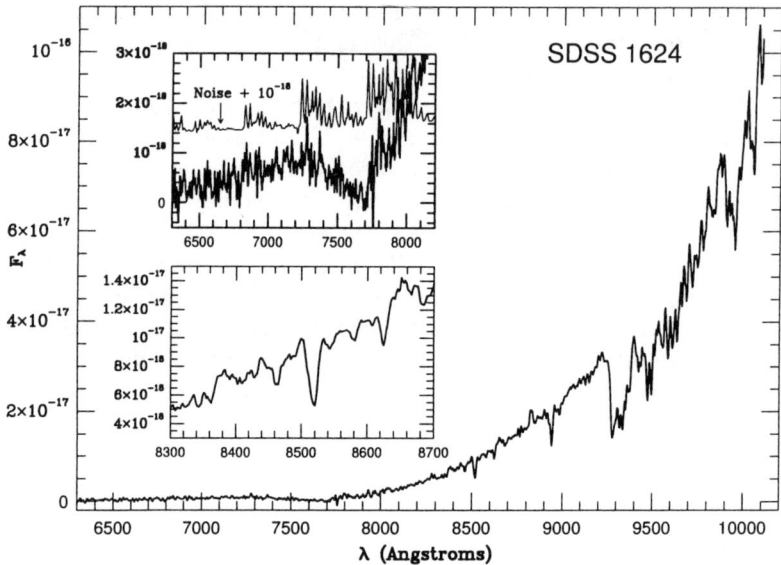

Fig. 2. The Keck II spectrum of 2M1624+0029, F_λ vs. $\lambda(\text{Å})$, boxcar-smoothed by 5 pixels (10Å). Two inset boxes highlight the "blue" and 8300-8700Å regions. The noise (variance) spectrum is shown in the top inset (shifted upward by 10^{-18}) and is flat over this interval.

whether the same will be possible for T dwarfs. At the very least, the T's are turning out to pose more problems. In Dr. Tsuji's paper, the L sequence represents the waxing of dust dormation, while the dust wanes among the T's. He points out that it is difficult to calculate from first principles the stability of dust in a "static" atmosphere. One of his concluding remarks is that "we should probably learn the method of meteorology in the future."

We all know what this remark means for the Earth's atmosphere – and for those of other planets and the Sun – which are studied both spatially and temporally. Indeed, all of these vary both spatially and temporally. It would be the height of naivete to expect anything less from brown dwarfs. One has to worry about non-static atmospheres being the reality. One dimensional atmospheric modelling will have its limitations.

Observers have started to look for "weather" on brown dwarfs. Tinney and Tolley [16] did a limited time series imaging sequence on two late M/L brown dwarf targets, using narrow band filters "on" and "off" a TiO band, to search for variability in T_{eff}, and/or dust condensation. The result may be considered exploratory, as a very comprehensive observing effort is needed to look for variations on, for example, the rotation time scale.

Bailer-Jones and Mundt [1] have published the most comprehensive study of photometric variations in late M to late L dwarfs. These authors find the best evidence for significant variations among the latest L objects. Since magnetic activity, based on Hα emission, appears to decline over this spectral interval [5],

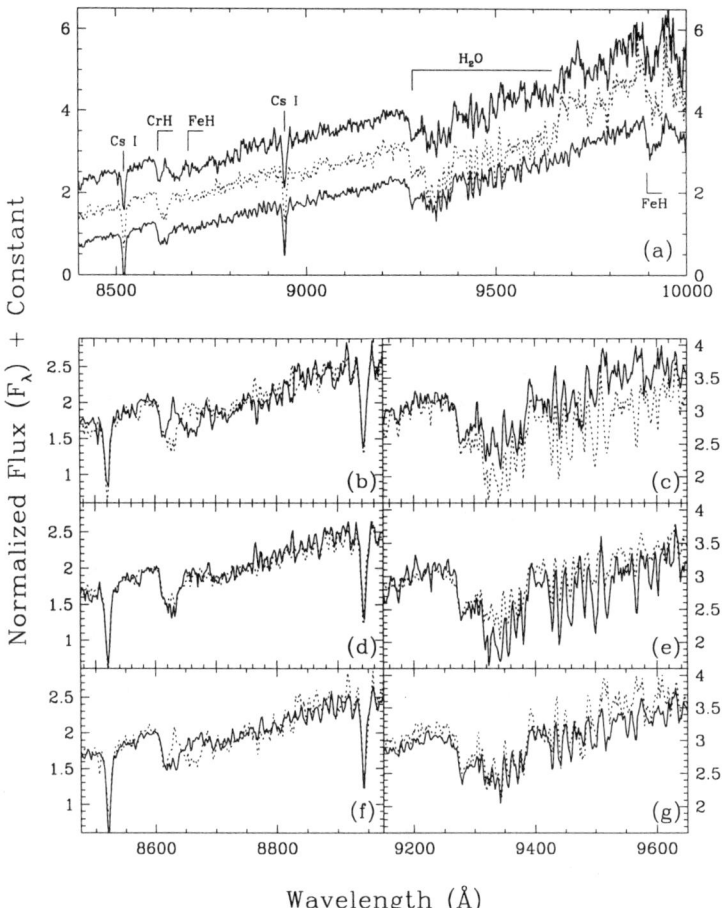

Fig. 3. Comparisons of the nightly spectra of the L8 V Gl 584C. a) The region between 8400 and 10000Å. b),d) and f) are pairings of two nights covering the CrH band. c), e) and g) are pairings covering a H$_2$O band, as discussed in Kirkpatrick et al. [8].

these authors consider it more likely that the photometric variations of late L dwarfs are due to variable condensates.

Interestingly enough, the 2MASS group has discovered spectrum variability in the red spectrum of one of the latest L8V dwarfs, the companion object Gl 584C. Shown in Fig. 3 are three spectra, to be discussed in detail in Kirkpatrick et al. [8]. Eleven total integrations were obtained on three nights, but the nightly sums are shown here. It is again not enough information for many conclusions to be drawn, though variability at 8600-8700Å longward of the CrH bandhead and in the H$_2$O band longward of 9300Å is apparent.

There are going to be two problems with brown dwarf weather observing projects. First, there will not be spatial resolution, for objects which we now know form clouds, for example. Secondly, telescope time assignment committees will not award enough time on a large telescope to observe these faint objects with sufficient time resolution to sample variability on the relevant time scales.

Acknowledgements

I thank my colleagues Neill Reid and Davy Kirkpatrick for figures shown here.

References

1. C.A.L. Bailer-Jones, and R. Mundt, A&A , in press, astro-ph/0012224 (2000)
2. G. Basri, S. Mohanty, F. Allard, P.H. Hauschildt, X. Delfosse, E.L. Martin, T. Forveille, and B. Goldman, ApJ **538**, 363 (2000)
3. M.S. Bessell, AJ **101**, 662 (1989)
4. A. Burrows, M.S. Marley, C.M. Sharp: ApJ **531**, 438 (2000).
5. J.E. Gizis, D.G. Monet, I.N. Reid, J.D. Kirkpatrick, J. Liebert, and R.J. Williams, AJ **120**, 1085
6. J.D. Kirkpatrick, I.N. Reid, J. Liebert, R. Cutri et al. : ApJ **519**, 802 (1999)
7. J.D. Kirkpatrick, I.N. Reid, J. Liebert, J.E. Gizis et al. : AJ **120**, 447 (2000)
8. J.D. Kirkpatrick, C.C. Dahn, D.G. Monet, I.N. Reid, J.E. Gizis, J. Liebert and A.J. Burgasser AJ , in press (2001)
9. J. Liebert, I.N. Reid, A. Burrows, A.J. Burgasser, J.D. Kirkpatrick, J.E. Gizis: ApJ **533**, 155 (2000)
10. J.B. Oke, J.G. Cohen, M. Carr, J. Cromer, J., A. Dingizian, F.H. Harris, S. Labreque, R. Lucinio, W. Schaal, H. Epps, and J. Miller, PASP , **107**, 375 (1995)
11. Ya. Pavlenko, M. R. Zapatero Osorio, and R. Rebolo, A&A , **355**, 245 (2000)
12. I.N. Reid, 2001, in *Galactic Structure, Stars, and the ISM*, ASP Conf. Series, in press (astro-ph/0010202)
13. I.N. Reid and J.E. Gizis AJ **114**, 1992 (1997)
14. I.N. Reid, J.E. Gizis, J.D. Kirkpatrick, and D.W. Koerner, AJ , **121**, 489 (2001)
15. M.A. Strauss et al. : ApJ **522**, L61 (1999)
16. C. Tinney and A.J. Tolley, MNRAS **304**, 119 (1999)
17. T. Tsuji, K. Ohnaka, and W. Aoki, A&A , **305**, L1 (1996)

Unified Model Photospheres for Ultracool Dwarfs of Types L and T

T. Tsuji

Institute of Astronomy, School of Science, The University of Tokyo,
2-21-1 Osawa, Mitaka, Tokyo, 181-0015 Japan

Abstract. The presence of the two distinct groups of ultracool dwarfs (UCDs), 'L' and 'T' types, is now well established: L dwarfs are thought to be dusty while T dwarfs show strong bands of methane (CH_4) but little evidence for dust in their spectra. So far, different model sequences, which we referred to as the dusty model (or case B) and dust-segregated model (or case C) have been considered for L and T dwarfs. We now propose instead that these two groups of UCDs may be understood as a temperature effect in a unique sequence of the model photospheres in which dust always exists but only in the restricted region (where $1800 \lesssim T \lesssim 2000\,\mathrm{K}$) referred to as an active dust zone (or as a dust cloud). This is a natural consequence of considering not only dust formation but also its segregation process in the photosphere. By this model sequence, the dust-column density in the observable photosphere first increases for cooler objects and the infrared colours become redder from late M to L dwarfs. On the other hand, the dust-column density in the observable photosphere decreases in the objects cooler than the latest L dwarfs as the active dust zone moves to the optically thick region deep in the photosphere and the infrared colours turn blueward towards the coolest T dwarfs. In this way, the observed colours and spectra of UCDs through L and T types can be explained consistently by a single grid of unified model photospheres. More generally, an important conclusion is that the photospheric dust formation is effective only in warmer deep regions. This is contrary to the general belief that dust forms in cooler surface regions.

1 Introduction

Progress in observations of ultracool dwarfs (UCDs) has been quite substantial even in the short time since the discoveries of the prototypes such as GD165B [1] and Gliese 229B [23]. A large number of UCDs similar to GD165B have been discovered by the Two-Micron All-Sky Survey (2MASS) [15] and the DEep Near-Infrared Sky (DENIS) survey [8]. A sample of these L-type dwarfs is already large enough that its sub-types from L0 to L8 have been defined by Kirkpatrick et al. [16] [17]. On the other hand, objects similar to Gliese 229B were more difficult to find, but it was not long before a dozen of cool brown dwarfs, referred to as methane dwarfs or T-type dwarfs, was discovered by the 2MASS [2] [3] [4] and the Sloan Digital Sky Survey (SDSS) [28] [35]. Finally, possible transition objects between L and T dwarfs were found by Leggett et al.[19] and they were classified as early T dwarfs. This discovery confirmed that the L- and T-types form a single spectral sequence and may not be representing any kind of bifurcation.

The new spectral types L and T are added to the spectral types of O, B, A, F, G, K, & M (with branching into R-N and S) established nearly a century ago.

While the spectral sequence from O to M types is well understood as a temperature sequence by ionisation and dissociation theory, the problem is why dust apparently disappears in the cooler T dwarfs even if the spectral types L and T can be understood as an extension of the M type to the lower temperatures. This article shows that the types L and T are indeed understood as a temperature sequence and that this is because the photospheric dust formation is effective only in the warmer deep region whose location depends on the effective temperature. This paper briefly reviews intriguing investigations that were carried out prior to the solution (Sect. 2) and discusses the resulting unified dusty model in some detail (Sect. 3). Then it is shown that that the observed infrared colours of UCDs may be explained consistently by our grid of unified models and, further, that the infrared two colour diagrams provide a useful constraint on the location of the dust zone in the photosphere (Sect. 4). Our models are also applied to a preliminary analysis of the spectra of UCDs (Sect. 5). While recognizing some inherent difficulties in probing dusty photospheres, we discuss some of the consequences of our new models (Sect. 6).

2 Modelling the Photosphere of Ultracool Dwarf Stars

Stellar photospheres are too hot for dust to form in general, but UCDs are exceptional in that the thermodynamical condition of condensation is well met in their photospheres. We had to confront how to treat dust in modelling stellar photospheres for the first time for which some trial and error were necessary. At first we considered dust formation and its segregation in different models (Sect. 2.1). Our initial attempt to combine them in a single model was applied to the specific case of Gliese 229B (Sect. 2.2). We now show that dust formation and its segregation process should be treated more consistently in a single photospheric model and we finally have the unified model photosphere of UCDs in which dust always exists but at different locations in L and T dwarfs. We hopefully conclude our exploratory stage of investigating dusty model photospheres within the framework of the classical theory of stellar photosphere (Sect. 2.3).

2.1 Dusty and Dust-Segregated Models: 1996 - 1998

It was recognized that the condition of condensation is well fulfilled in the photospheres of cool dwarfs [32], but it was unknown how dust forms in the photospheric environment. A problem is when nucleation begins after the supersaturation ratio $S = p/p_{sat}$ (p_{sat} is the saturation vapour pressure) exceeds unity. We assumed two extreme cases: One is a dust-free model in which dust is not formed even if $S > 1$ and the other is a dusty model in which dust forms as soon as S exceeds unity. It is found that the dusty models explain the observed characteristics of late M dwarfs [32] [13] as well as L-type prototype GD 165B [33]. In this case, we assume that the small dust grains formed at relatively high temperatures remain well mixed with the gaseous components. On the other hand, the genuine brown dwarf Gliese 229B discovered by Nakajima et al. [23] shows little evidence for dust but could be explained rather well by our dust-free model developed before the discovery of the brown dwarf [31]. We interpreted this result as due to the segregation of dust which once formed and grew too large to be sustained in the photosphere of the cool brown dwarf [33].

The preceding results are easily consistent with the classical nucleation theory according to which the dust growth cannot start before its radius r_{gr} reaches the critical radius r_{cr}, where the Gibbs free energy of condensation shows the maximum. The dust grains with $r_{gr} < r_{cr}$ are thermodynamically unstable, in the sense that the dust grains formed will soon dissolve and *vice versa*. In other words, such small dust grains are in detailed balance with the gaseous mixture and hence can easily be sustained in the photosphere. For this reason, such small grains that failed to be the stable large grains, play an important role as sources of opacity in the photosphere. On the other hand, the dust grains with $r_{gr} > r_{cr}$ are stable and grow larger and larger. Such large grains, however, will segregate from the gaseous mixture and no longer be sustained in the photosphere. Thus, large grains may be formed, but they may precipitate below the photosphere and will play little role as sources of opacity. Then, we distinguished three cases of $r_{gr} = 0$, $r_{gr} \lesssim r_{cr}$ and $r_{gr} > r_{cr}$ which we referred to as case A (supersaturated case), B (dusty case) and C (dust-segregated case), respectively [30]. We hoped that L and T dwarfs could be accounted for by the dusty models (case B) and dust-gas segregated models (case C), respectively. One difficulty, however, was the large excess of the optical flux predicted by the dust-segregated model compared with the observation of Gliese 229B [10].

2.2 A Hybrid Model: 1999

An initial motivation to consider a hybrid model, which consists of the warm dust in the deeper layer and cool volatile molecules in the upper layer, was to explain a large flux depression in the optical spectrum of Gliese 229B [34]. In fact, if dust plays some roles in depressing the optical flux, only the dust deep in the photosphere will work for this purpose, since the dust in the surface region will mask other prominent spectral features such as due to CH_4 and H_2O. At the same time, however, it was noticed that the effect of the pressure-broadened wings of alkali metals depresses the optical flux significantly [34] [6]. Therefore dust is not necessarily called for to explain the optical flux depression.

Nevertheless, we will show that the warm dust deep in the photosphere will have noticeable observable effects in L and early T dwarfs (Sect. 2.3). In fact, the significance of the hybrid model is not necessarily to explain the optical flux depression, but rather it involves an important idea to be developed to a unified model of UCDs. Our previous dusty (case B) or dust-segregated (case C) models were based on the assumptions that the dust once formed remains throughout the photosphere in L dwarfs or segregation process takes place throughout the photosphere in T dwarfs, respectively. But, it is difficult to understand why the different cases are realized for the same physical conditions that may be found somewhere in the photospheres of L and T dwarfs. Although these cases B and C models could explain some characteristics of a few selected M, L and T dwarfs, new observations on a larger sample revealed some inconsistencies and suggested a more realistic model somewhere in between these two extreme cases [29]. Our hybrid model may already be suggesting a way to relax these issues.

2.3 The Unified Model: 2000

Once again, we remember how dust forms in photospheric conditions. Dust forms as soon as temperature is lower than the condensation temperature (T_{cond}), but the dust will soon grow too large at slightly lower temperature, say T_{cr} (critical temperature), when the dust size reaches its critical radius r_{cr}. Thus, in the region with $T \lesssim T_{cr}$ in the photosphere, dust will be large enough to segregate from the gaseous mixture and soon precipitate below the photosphere. Only in the region with $T_{cr} \lesssim T \lesssim T_{cond}$, the dust grains will be small enough ($r_{gr} \lesssim r_{cr}$) to be sustained in the photosphere and it is such small dust grains that play an important role as opacity sources. We refer to this region with $T_{cr} \lesssim T \lesssim T_{cond}$ as an active dust zone. Thus, the dust effectively exists only in the relatively warm region deep in the photosphere, and this means that a dust layer or a cloud is formed in the photosphere. This is a natural consequence of consistently considering not only dust formation but also its segregation process. In this way, a model with the warm dust layer deep in the photosphere can be constructed anew based on a clear physical basis rather than as a "hybrid" of any preceding models as its constituents. For this reason, it is not appropriate to refer to such a model with the active dust zone as a hybrid model and may be referred to as a unified model to distinguish it from the previous models.

Now, a major problem is to find the temperatures that define the active dust zone. The condensation temperature T_{cond} is easily found from the thermochemical computation during the model construction to be $T_{cond} \approx 2000$ K for corundum and iron, which first form in the photospheres of UCDs [32]. On the other hand, the critical temperature T_{cr} is more difficult to find. This should in principle be determined from the detailed analysis of the dust-gas segregation process, but it still seems to be premature to solve this problem theoretically. Instead, we treat T_{cr} as a free parameter to be found empirically. Actually, we find that T_{cr} can be constrained well by the infrared colours which show red limits at late L dwarfs [17] and we find it to be $T_{cr} \approx 1800$ K (Sect. 4). The critical radius r_{cr} itself is more difficult to estimate, but astronomical grains of 0.01 μm or smaller are known and we may assume that $r_{cr} < 0.01\,\mu$m. On the other hand, mass absorption coefficients of dust grains depend little on the grain sizes so long as the grains are smaller than about 0.01 μm, and the value of r_{cr} gives little direct effect on our actual modelling.

It is to be noted that the active dust zone can be found in all the cool dwarfs with $T_{eff} \lesssim 3000$ K. For objects with very low T_{eff} near 1000 K, this active dust zone is situated too deep in the photosphere (where $\tau_{Ross} > 1$, since $T \approx T_{eff}$ at $\tau_{Ross} \approx 1$) and this should be the reason why dust apparently shows little observable effects in cool T dwarfs. On the other hand, for the relatively warm objects with T_{eff} above about 1500 K (see Sect. 4), the active dust zone is situated nearer to the surface ($\tau_{Ross} < 1$) and this explains why L dwarfs, whose T_{eff} may be higher than about 1500 K [26], appear to be dimmed by dust. The case of the early T dwarfs recently discovered [19], the active dust zone may just be situated near the optical depth unity and thus these objects with $1200 \lesssim T_{\text{eff}} \lesssim 1400$ K may represent the L/T transition objects.

3 Physical Structures of the Unified Models

We will discuss some details of our unified model for the case of $T_{\text{eff}} = 1300$ K as an example in Fig. 1. The active dust zone is shown by the dotted area, the upper boundary of which is defined by the condensation line of corundum (Al_2O_3) shown by the dashed line. Before the active dust zone terminates at $T \approx T_{\text{cr}} \approx 1800$ K, iron (Fe) condenses at its condensation line. However, enstatite ($MgSiO_3$) may form outside the active dust zone and this means that enstatite will segregate as soon as it is formed. This may be possible, since enstatite will easily form with corundum and/or iron as the seed nuclei and grow rapidly. By the same reason, other solid species that may form at lower temperatures will precipitate as soon as they are formed. For this reason, only the dust species formed at relatively high temperatures above about $T_{\text{cr}} \approx 1800$ K work as the active dust (*i.e.* as source of opacity) and hence give significant effect on the photospheric structure. This fact may simplify the construction of models since it is enough to consider only the high temperature condensates

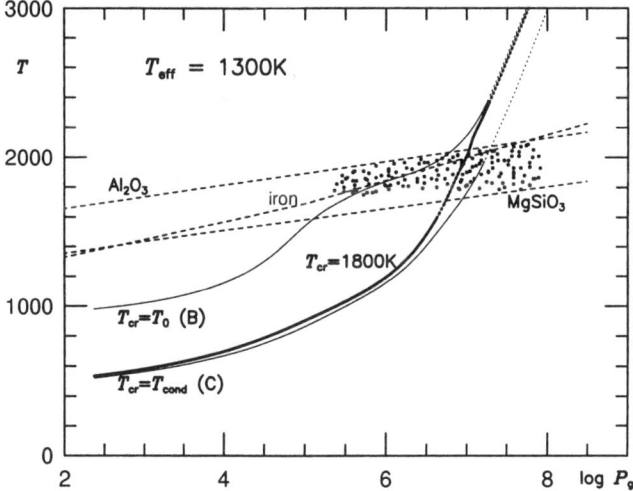

Fig. 1. Model photospheres of $T_{\text{eff}} = 1300$ K (solar metallicity, $\log g = 5.0$ & $v_{\text{micro}} = 1 \,\text{km}\,\text{s}^{-1}$). The heavy line represents the unified model with $T_{\text{cr}} = 1800$ K while the thin lines the limiting cases of $T_{\text{cr}} = T_0$ (case B) and $T_{\text{cr}} = T_{\text{cond}}$ (case C). The solid and dotted lines represent radiative and convective zones, respectively. The condensation lines of corundum (Al_2O_3), iron and silicate ($MgSiO_3$) are shown by the dashed lines. The active dust zone is indicated by the dotted area

such as corundum and iron as sources of opacity. In the active dust zone, the temperature gradient is quite steep because of the high opacity of the dust and the model is convectively unstable near T_{cr}. For this reason, our unified model shows the outer and inner convective zones separated by an intermediate or detached radiative zone (Fig. 1), as discussed in the case of the hybrid model for Gliese 229B [34].

In Fig. 1, the resulting structure of our new model is found between the previous dusty (cases B) and dust-segregated (case C) models as can be expected. In fact, we assumed that the dust grains once formed remain small enough throughout the photosphere and never precipitate up to the stellar surface in our previous dusty model (case B). This is equivalent to have assumed $T_{cr} = T_0$ (T_0: surface temperature). On the other hand, we assumed that the dust grains will precipitate as soon as they are formed in our previous dust-segregated model (case C) and this is equivalent to have assumed $T_{cr} = T_{cond}$. Thus, our previous models of cases B and C represent the extreme limiting cases of our new model. The active dust in our new model works to heat up the photosphere near the active dust zone but not so much as in our previous case B model in the upper layer. On the other hand, volatile molecules work as dominant sources of opacity after dust has precipitated in the layer above $T = T_{cr}$ in our new model and the photosphere is cooled appreciably by the cooling effect of the volatile molecules as in our case C model. It is to be noted, however, that the photospheric structure of the unified model approaches that of the previous case B in the region below the active dust zone.

4 Colours

The infrared colours of UCDs are not necessarily redder for cooler objects but turn to blue in T dwarfs after passing the red limits around late L dwarfs [17]. These observations indicate that there should be an additional parameter other than T_{eff} in determining the colours. This parameter should be related to the dust in the photosphere and we identify it with T_{cr}. We will show that the rather complicated behaviours (Figs. 2 & 3) of the infrared colours are well understood by our unified models characterized by T_{eff} as well as by T_{cr} and that the effect of these two parameters can be separated on the infrared two-colour diagrams.

4.1 $(J - H, H - K_s)$ Diagram

The $J - H$ and $H - K_s$ are the most observed colours for a large sample of UCDs. We reproduce the observed colours of M, L and T dwarfs including the early T-types, by open triangles, circles and squares, respectively, in Fig. 2. With the discovery of the early T dwarfs [19], which may be the L/T transition objects, it now appears that the UCDs show a continuous loop counter-clockwise on the $(J - H, H - K_s)$ diagram from M to T dwarfs via L dwarfs. Also, an interesting feature is the presence of the red limits for the infrared colours [17].

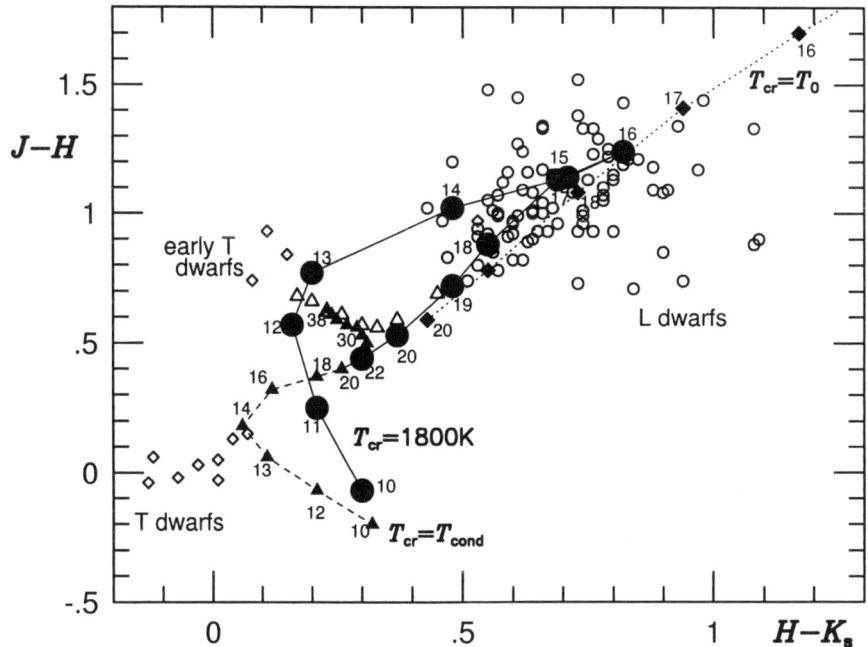

Fig. 2. ($J - H$, $H - K_s$) diagram. Observed colours of M, L and T dwarfs are shown by the open triangles [18], circles [16] [17] and squares [22] [28] [2] [3] [4] [35] [19], respectively. Predicted colours based on our unified models with $T_{cr} \approx 1800$ K are shown by the filled circles while those of the dusty (case B) and dust-segregated (case C) models by the filled squares and triangles, respectively. The numbers attached are T_{eff}'s in units of one hundred Kelvin (the steps of T_{eff}'s are 100 or 200 K)

In Fig. 2, the predicted colours based on our previous case B models (filled squares) could explain the very red colours of some L dwarfs, but could not the presence of the observed red limits of $J - H$ and $H - K_s$. On the other hand, our case C models (filled triangles) could explain the very blue colours of T dwarfs (but predicted $H - K_s$'s are too red by about 0.3 mag. as compared with observed and this is due to the difficulty to predict the K flux accurately by our models as will be noted in Sect. 5.2), but could not the reddening of $J - H$ towards early T dwarfs. On the other hand, our new model (filled circles) roughly explains the general trend of the observed colours through M, L, L/T and T dwarfs by a single grid of the model photospheres with $T_{cr} \approx 1800$ K.

The redness of the infrared colours is essentially determined by the mass-column density of the active dust in the observable photosphere. Since the active dust zone is within the optically thin regime in the relatively warm objects, the dust-column density in the observable photosphere first increases towards cooler objects from late M to L dwarfs and the infrared colours show reddening in agreement with observations. However, the dust-column density in the observable photosphere decreases towards the coolest objects even if the mass-column

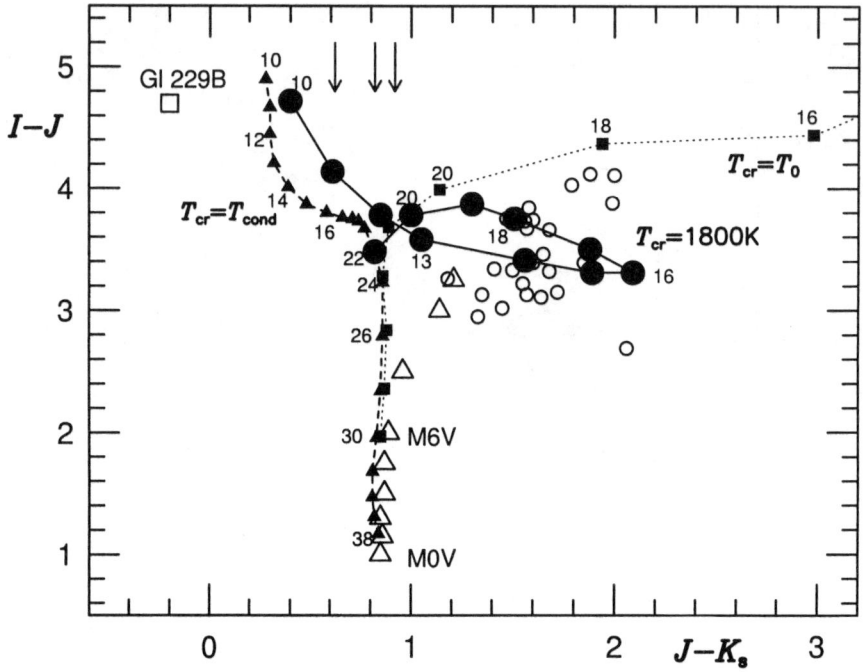

Fig. 3. $(I-J, J-K_s)$ diagram. The arrows indicate the values of $J-K_s$ for three L/T transition objects [19]. See Fig. 2 legend for details

density of the active dust zone itself increases, because it now penetrates into the optically thick regime in the cooler objects. Our model grid predicts that this takes place at $T_{\rm eff} \approx 1600\,{\rm K}$ (Fig. 2), which may correspond to the latest L dwarf [26]. This explains the presence of the red limits in the $(J-H, H-K_s)$ colours and the bluer colours of L/T transition objects as well as of T dwarfs.

The observed red limits of $(J-H, H-K_s) \approx (1.3, 0.8)$ [17] are well explained by our model grid based on $T_{\rm cr} = 1800\,{\rm K}$ which predicts the red limits of $(J-H, H-K_s) \approx (1.2, 0.8)$ at $T_{\rm eff} \approx 1600\,{\rm K}$ (Fig. 2). The mass-column density of the active dust zone in the observable photosphere should also be larger for the lower $T_{\rm cr}$, since this means that the active dust zone extends outward and the red limits will still be redder. In fact, our grid based on $T_{\rm cr} = 1600\,{\rm K}$ predicts the red limits of $(J-H, H-K_s) \approx (1.5, 1.1)$. On the other hand, another grid based on the higher $T_{\rm cr} = 1900\,{\rm K}$ predicts the red limits of $(J-H, H-K_s) \approx (1.0, 0.5)$. Thus, the value of $T_{\rm cr} = 1800\,{\rm K}$ can be regarded as being well constrained.

4.2 $(I-J, J-K_s)$ Diagram

We show the observed and predicted $I-J$ and $J-K_s$ colours in Fig. 3. Here, the observations show bifurcation to the red (L dwarfs: open circles) and blue (Gliese 229B: open squares) sequences. The bifurcation could apparently be explained by our previous case B (filled squares) and C (filled triangles) models. However, our unified models (filled circles) indicate a possibility that these sequences are in fact understood as a single sequence. The observed $J - K_s$ colours of the three L/T transition objects [19] shown by the arrows in Fig. 3 confirm that the observed colours form a single continuous loop in the $(I - J, J - K_s)$ diagram. The reason for this is essentially the same as for $(J-H, H-K_s)$ figure. Also, the observed red limit of $J - K_s \approx 2.1$ is well reproduced by our model grid based on $T_{cr} = 1800$ K (Fig. 3), while it could not be predicted at all by our previous cases B and C models. For comparison, the predicted red limits are $J-K_s \approx 2.6$ and 1.5 for our grids based on $T_{cr} = 1600$ and 1900 K, respectively. Again, our choice of $T_{cr} = 1800$ K is well justified by the $(I - J, J - K_s)$ diagram. However, our predicted $J - K_s$'s are too blue for the late M and early L dwarfs while too red for the coolest T dwarf Gliese 229B (also see Fig. 5). Probably, something may be still missing in our opacity data.

5 Spectra

In our unified models, only the dust in the active dust zone located within the observable photosphere gives some observable effects, but even such a small amount of dust gives appreciable effects on spectra as well as on colours because of the very large extinction of dust. We first show a general characteristics of the infrared spectra based on our new grid of the unified models (Sect. 4.1). Then, we rediscuss the prototype of T-type Gliese 229B (Sect. 4.2) and that of L-type GD165B (Sect. 4.3) by our models.

5.1 Predicted Spectra of the Unified Models

The predicted spectra (in F_ν unit) based on our new models shown in Fig. 4 can be seen to be well consistent with the infrared colours discussed in Sect.4. For example, the mass-column density of the active dust zone in the observable photosphere is increasing in our model of $T_{\rm eff} = 1800$ K (Fig. 4a) and the effect of dust on the spectrum is appreciable. The effect of dust extinction on the J band is largest in case B, shows no effect in case C and it is just intermediate between these extreme cases in our new model, as is the $J - H$ (Fig. 2). At the same time, the dust also contributes to heat the photosphere and the K band region shows the opposite tendency because the effect of the H_2 collision-induced absorption (CIA) which is less important in the warmer photosphere of the larger dust-column density. In the case of $T_{\rm eff} = 1300$ K (Fig. 4b), the mass-column density of the active dust zone itself still increases, but the part in the observable photosphere decreases since part of it is now below the optical depth unity. For this reason, the effect of dust is only modest, but the J flux still suffers considerable extinction by the dust resulting in the reddening of $J - H$ (Fig. 2). Finally, in the model of $T_{\rm eff} = 1000$ K (Fig. 4c), the predicted infrared spectrum based on our new model differs little from that of the case C, as are

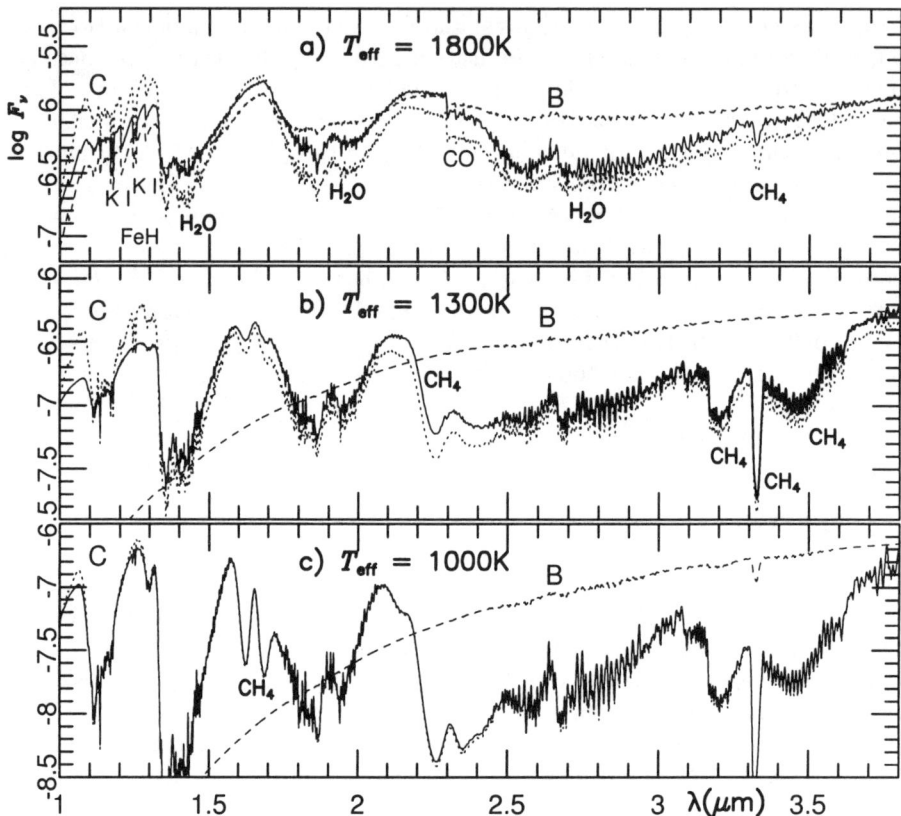

Fig. 4. The predicted spectra (in erg/cm^2/sec/Hz) based on our models ($T_{cr} = 1800$ K) are shown by the solid lines while those of the limiting cases of $T_{cr} = T_0$ (dusty case B) and $T_{cr} = T_{cond}$ (dust-segregated case C) by the dashed and dotted lines, respectively. (a) $T_{eff} = 1800$ K. (b) $T_{eff} = 1300$ K. (c) $T_{eff} = 1000$ K

the infrared colours (Figs. 2 & 3). This is because the active dust zone is situated below the observable photosphere of this very cool model.

It is interesting to see in our unified model that the Q branch of the CH$_4$ ν_3 fundamentals appears by $T_{eff} = 1800$ K and that the weaker combination bands at 1.6 and 2.2 μm are strong by $T_{eff} = 1300$ K. These results find observational support in the recent detections of CH$_4$ bands in L dwarfs reported in this meeting by Geballe and by Noll as well as in the early T dwarfs by Leggett et al. [19]. The CH$_4$ ν_3 fundamentals appeared at $T_{eff} = 1800$ K in our previous model C (Fig. 4 in [30] in which CH$_4$ bands were probably overestimated by the use of the smeared out CH$_4$ opacity), but we did not think this to be serious since we thought that our previous model C cannot be applied to L dwarfs. This conclusion remains unchanged even if CH$_4$ bands can be predicted by this model. However, our previous dusty model (case B) never predicts the CH$_4$

bands (Fig. 4a) and the recent detection of CH_4 in L dwarfs completely ruled out the possibility of the simple dusty models for L dwarfs. Thus, our new model provides a distinct possibility to explain the presence of methane as well as of dust in L dwarfs consistently.

In the present work, we use the line databases HITEMP [27] for H_2O and GEISA [11] for CH_4 ν_3 fundamentals, but we still use the smeared-out band models for CH_4 combination bands as well as for FeH. The present CH_4 linelist, however, is valid only at low-temperatures and its extension to the higher temperatures is urgently needed.

5.2 The Spectrum of the T Dwarf Prototype Gliese 229B

We compare the observed spectrum of Gliese 229B by Geballe et al. [9] (calibrated by Leggett et al. [20]) and by Oppenheimer et al. [25], with the predicted ones based on our new model of $T_{\text{eff}} = 1000\,\text{K}$ in Fig. 5. To show the effect of T_{cr}, we show two models: One with $T_{\text{cr}} = 1800\,\text{K}$ which we now believe to be the best (see Sect.4) and the other with $T_{\text{cr}} = 1600\,\text{K}$ which is close to the value we applied to Gliese 229B in our previous analysis [34]. Inspection of Fig. 5 reveals that the case of $T_{\text{cr}} = 1800\,\text{K}$ gives an overall better fit than the case of $T_{\text{cr}} = 1600\,\text{K}$. This fact confirms that our choice of $T_{\text{cr}} = 1800\,\text{K}$ is acceptable for the coolest T dwarfs as well. But we identify two major discrepant regions. First, the predicted flux appears to be higher than the observed at the K band region and this is the reason why the predicted $H - K_s$ and $J - K_s$ are too red (Sect. 4). Here, our opacity data (e.g. CH_4) may not be perfect and some unknown sources may also be possible at the very low temperatures. Second, the optical flux cannot yet be explained quantitatively and one possibility may be to improve the broadening theory of strong alkali metal lines as suggested by Burrows et al. [6].

Our previous choice of $T_{\text{cr}} = 1550\,\text{K}$ [34] was largely biased towards explaining the observed spectrum of Gliese 229B in the optical region, for which the predicted flux based on $T_{\text{cr}} = 1600\,\text{K}$ in fact shows a better fit (e.g. i-flux in Fig. 5). However, we now recognize that T_{cr} cannot be determined from such a very cool object alone whose active dust zone gives little observable effect. In fact, our new model with $T_{\text{cr}} = 1800\,\text{K}$ predicts almost the same spectrum as our previous dust-segregated model for $T_{\text{eff}} = 1000\,\text{K}$ (Fig. 4c). For this reason, it may not be possible to prove nor to disprove the presence of the warm dust deep in the photosphere by the analysis such as done on the spectrum of SDSS 1624 [21] [24]. Thus, the presence of the warm dust proposed for Gliese 229B [34] cannot be confirmed by this object itself, but can be deemed as well established by our analysis of a larger sample of UCDs (e.g. Sect.4).

5.3 The Spectrum of the L Dwarf Prototype GD 165B

Our old dusty model (case B) already explained the observed spectrum of GD165B rather well while our dust-segregated model (case C) could not [30] and GD165B may in fact be remembered as the first object in which the presence of dust in

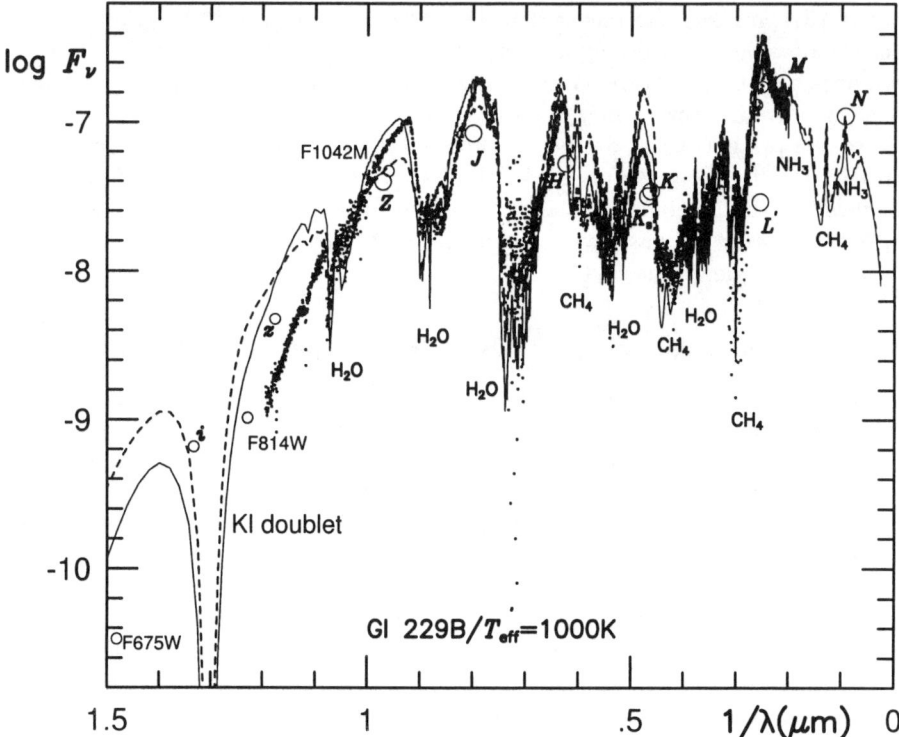

Fig. 5. Observed spectrum [9] [20] [25] and photometry data [22] [10] of Gliese 229B are shown by the dots and open circles, respectively. The predicted spectra (in erg/cm^2/sec/Hz) based on our unified models of $T_{\text{eff}} = 1000$ K are shown by the solid and dashed lines for the cases with $T_{\text{cr}} = 1800$ K and 1600 K, respectively

the photosphere was recognized [33]. But this result should be somewhat fortuitous in view of our new models and we rediscuss the observed spectrum of GD165B by Jones et al. [12] in Fig. 6. It is clear that our new models explain the observed spectrum as well, especially if we assume $T_{\text{eff}} = 1750$ K. It is to be remembered that the best fit with our previous dusty models was obtained for $T_{\text{eff}} = 1800$ K [33] and another independent analysis suggested $T_{\text{eff}} = 1900$ K [14]. Thus, the effect of the new model is to lower the estimated T_{eff}. Although all these models applied so far to GD165B provide more or less similar good fits, our new model is physically more reasonable and, moreover, our analyses of the infrared colours for a large sample of UCDs provides definite evidence for our new models (Sect. 4). Thus, GD165B may have $T_{\text{eff}} \approx 1750$ K and now be closer to the substellar regime although its substellar nature may still depend on the details of the evolutionary models [5] [7]. This result, however, may not be completely free from the difficulty to be discussed in Sect. 6.3, even though T_{cr} is relatively well estimated for the L dwarf regime.

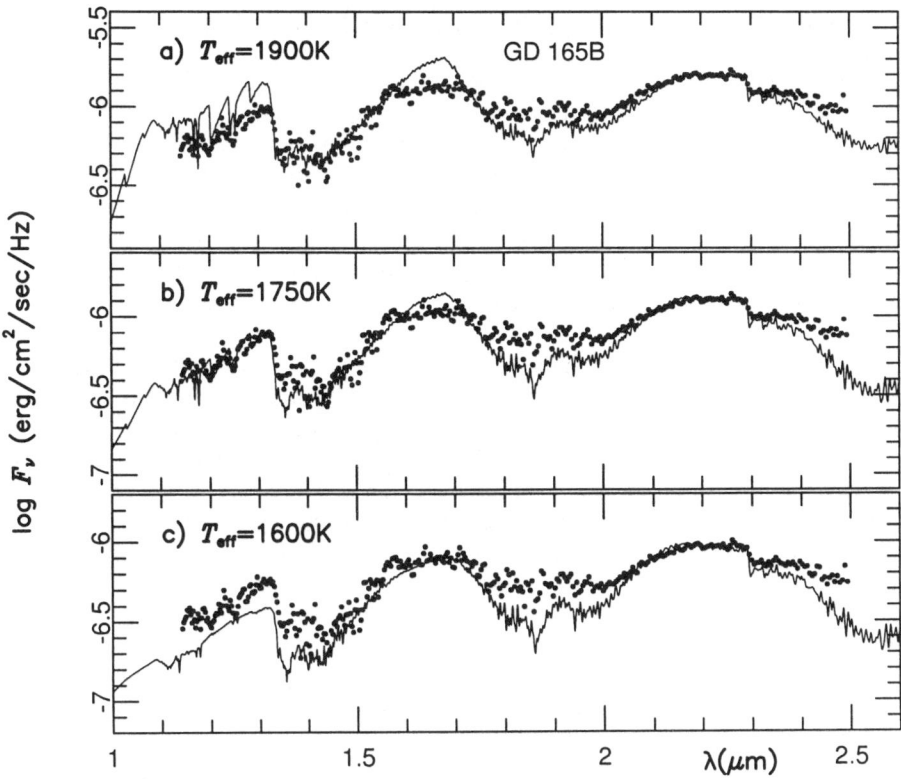

Fig. 6. Observed spectrum of GD 165B [12] shown by the dots is fitted (first at the K band region) to the predicted ones based on our unified models ($T_{cr} = 1800\,\mathrm{K}$) shown by the solid lines. (a) $T_{\mathrm{eff}} = 1900\,\mathrm{K}$. (b) $T_{\mathrm{eff}} = 1750\,\mathrm{K}$. (c) $T_{\mathrm{eff}} = 1600\,\mathrm{K}$

6 Discussion

An important conclusion on dust formation in the photospheric environment is that it is effective only in the warmer deep region and not in the cooler surface region (Sect. 6.1). Once this simple principle is realized, we can construct reasonably realistic model photospheres for UCDs and the new spectral types L and T can consistently be interpreted as a temperature sequence (Sect. 6.2). However, the details of spectra depend on the dust-column density in the observable photosphere and fully quantitative analyses of the spectra of dusty photospheres should have some inherent difficulties. Furthermore, it is difficult to prove the presence of dust if it is below the observable photosphere as spectroscopic diagnosis is impossible for anything below the photosphere (Sect. 6.3).

6.1 Dust Formation in the Photospheric Environment

Unlike the case of cool giant stars where dust forms in the outflow, dust in cool dwarf stars forms in the static photosphere and there should be fundamental differences in the dust formation mechanisms in low and high luminosity stars. One problem is how dust could be sustained in the photosphere of dwarf stars for a long time. After considering not only dust formation but also its segregation process, we arrive at a conclusion that dust formed can be sustained only in the region near the condensation temperatures in the photosphere (Sect. 2.3). This means that dust effectively exists in the rather warm region relatively deep in the photosphere contrary to the general belief that dust is more abundant in the cooler surface region. This conclusion may apply not only to UCDs but also to the dust formation in the photospheric environment in general, including extrasolar giant planets (hot Jupiters), proto stars, accretion disks etc.

This conclusion on photospheric dust formation is confirmed by the fact that our new models based on this assumption explain the observed colours (Sect.4) and spectra (Sect.5) rather well. On the other hand, our previous dusty (case B) and dust-segregated (case C) models, which represent the extreme limiting cases of $T_{cr} = T_0$ (T_0: surface temperature) and $T_{cr} = T_{cond}$, respectively, could not explain the observed colours of UCDs consistently (Figs 2 & 3). Thus, it is clear that our previous models are not realistic enough to be used for interpreting observed data and should no longer be used. The model photospheres of UCDs by other authors are also subject to the same criticism. For example, the DUSTY and COND models discussed recently by Chabrier et al. [7] correspond to the limiting cases of $T_{cr} = T_0$ and $T_{cr} = T_{cond}$, respectively. The photospheric models of cool brown dwarfs including giant planets by Burrows et al. [5] also essentially assumed $T_{cr} = T_{cond}$ throughout. In the case of the coolest brown dwarfs, where dust exists deep in the photosphere, the models with $T_{cr} = T_{cond}$ give essentially the same emergent spectra as the unified models. However, dust deep in the photosphere gives considerable effect on the structure of the inner photospheres and hence on the boundary condition for the interior models.

6.2 Stellar Spectral Classification Extended to UCDs

The spectral classification of L dwarfs by Kirkpatrick et al. [16] [17] is based on a large sample of the far-red spectra of UCDs (0.63 – 1.01 μm). The resulting spectral subclasses show good correlations with the infrared colours, although the correlations are not monotonic especially if T dwarfs are considered, but show a red limit at the latest L dwarfs [16] [17]. We have shown that the infrared colours of UCDs are essentially controlled by the mass-column density of the dust in the observable photosphere which first increases towards lower T_{eff} but shows a maximum at about $T_{eff} \approx 1600\,\mathrm{K}$ and then decreases towards the coolest T dwarfs (Sect. 4). Thus, the infrared colours are well correlated with stellar temperatures and this fact confirms that the spectral classification by Kirkpatrick et al. [16] [17] also represents the temperature sequence. It is remarkable that the spectral classification by Kirkpatrick et al. [16] [17] done on a purely empirical basis reflects the photospheric structure of UCDs so well. We also propose that L and T types represent the objects in which the dust-column density in the observable photosphere is increasing and decreasing respectively. Since dust is the major ingredient in determining the photospheric structure, this interpretation of L and T types should be more fundamental than the use of methane which is now observed both in L and T dwarfs (Sect. 5.1). Anyhow, it may be reasonable to have divided UCDs into L and T types.

So far, we have not as yet analysed directly the spectral features used as classification criteria [16] [17], including oxides such as TiO and VO, hydrides such as FeH and CrH and neutral alkali metals. The absorption bands due to the refractory compounds should be formed below the active dust zone where refractory elements are not yet depleted in dust. This dust free zone in the observable photosphere (*i.e.* in $\tau < 1$) shrinks as the active dust zone moves towards deeper region in cooler L dwarfs and this is one reason why TiO and VO are weaker in cooler L dwarfs. Further, the mass-column density of the active dust above this molecule-dominated region is larger for cooler L dwarfs and molecular bands suffer larger extinction by the dust. Besides these two major effects of dust, the observed band strengths also depend on the gas phase chemical equilibrium. For example, TiO and VO attain their maximum abundances already in late M dwarfs while hydrides such as FeH and CrH may still be increasing in the L dwarf regime because of their lower dissociation energies. For this reason, hydrides are well observed in L dwarfs. On the contrary, alkali metals may be abundant in the region above the active dust zone and may be stronger for cooler objects including T dwarfs. Thus, atomic and molecular spectra will provide abundant information on the structure of the dusty photosphere.

The spectral sequence of O – M can be understood as a temperature sequence by considering ionisation and dissociation in gaseous mixture. The new spectral types L and T can be understood as a temperature sequence in which the dust forming region moves from the optically-thin region in L dwarfs to the optically-thick region in T dwarfs. Thus, the stellar spectral classification including the L- and T-types can be understand as a single sequence of temperature and a large variety of spectra from OB stars to brown dwarfs including L and T types can be interpreted by a simple thermodynamics including dust condensation and segregation in addition to ionisation and dissociation.

6.3 Spectroscopic Diagnosis of Dusty Photospheres

The observed colours and spectra of UCDs can in principle be interpreted by our unified models, but this does not imply that the quantitative spectroscopy of UCDs with the same accuracy as in non-dusty stars can be possible. The inherent difficulty is that the spectra of dusty objects depend strongly on the dust-column density in the observable photosphere (Sect. 5). Since dust shows no observable feature by itself, it is very difficult to determine the dust-column density directly. We can estimate it based on our models once we know $T_{\rm cr}$ which is estimated empirically by the observed colours (Sect. 4). Similar empirical approaches based on other observables such as atomic and molecular spectra may be tried, but a problem is how to separate the effects of T_{cr} and T_{eff} on the observables. Also, beside such empirical estimations, T_{cr} may hopefully be determined by the analysis of the detailed processes of dust growth and its segregation coupled with the dynamical processes of the meteorological scale. At present, however, such an *ab-initio* approach seems to be more difficult and it is not sure if it provides a more accurate estimation of T_{cr} than the empirical estimation.

Another problem in our empirical approach is that we assumed T_{cr} to be the same for all the models, but T_{cr} should certainly depend somewhat on T_{eff}. In fact, our models based on $T_{cr} \approx 1800\,\mathrm{K}$ and extended to T_{eff} above $2000\,\mathrm{K}$ showed only minor effect of dust and may fail to explain the observed spectra of late M dwarfs which are already known to show the effect of dust [32] [13]. Probably, our estimate of $T_{cr} \approx 1800\,\mathrm{K}$ may be valid for L dwarfs from which this result was obtained. We hope, however, that essentially the same approach can be possible to the model photospheres of late M dwarfs with somewhat lower value of T_{cr}. On the contrary, the exact value of T_{cr} may not be important in the coolest T dwarfs in which the active dust zone is below the observable photosphere, but information on dust will be almost lost from the spectra.

7 Concluding Remarks

We believe that we have shown a possibility for the unified model photospheres of UCDs including L and T dwarfs and have finally found an empirical approach to consistently treat dust formation and its segregation process in the photospheric environment. A natural consequence of our approach is that dust exists only in the restricted region deep in the photosphere, and thus a warm dust layer or a cloud is formed in the photosphere. Our approach is based on a simple assumption that only small dust grains can be sustained in the photosphere, and we show that a self-consistent non-grey model photosphere can be developed without any other ad-hoc assumption. However, once dust appears in the photosphere, it introduces some inherent difficulties. One problem is that the gas-dust phase change cannot be treated properly within the framework of the classical theory of stellar atmosphere and in future we should probably learn the recipes of meteorology. For example, we said nothing about the fate of the dust grains precipitated below the photosphere, where they may evaporate. These large grains may give some effects on the photospheric structure as well as on the observable properties of dusty objects. Also, unlike atoms and molecules that show well defined spectra, dust shows no identifiable spectrum especially in the case of UCDs and a formidable problem is how to know the dust-column density in the observable photosphere when we apply our models to the fully quantitative analysis of observed spectra of UCDs. It is desirable to apply model photospheres with these limitations in mind, even though model photospheres can be well useful as a guide for interpretation and analysis of the observed data.

Acknowledgements

I thank Tadashi Nakajima for helpful discussion throughout this work and Hugh Jones for careful reading of the text with useful comments. I also thank Tom Geballe, Hugh Jones and Ben Oppenheimer for making available their spectra in digital form.

References

1. E. E. Becklin, B. A. Zuckerman: Nature **336**, 656 (1988)
2. A. J. Burgasser, J. D. Kirkpatrick, M. E. Brown et al.: ApJ **522**, L65 (1999)
3. A. J. Burgasser, J. D. Kirkpatrick, R. M. Cutri et al.: ApJ **531**, L57 (2000)
4. A. J. Burgasser, J. C. Wilson, J. D. Kirkpatrick et al.: AJ **120**, 1100 (2000)
5. A. Burrows, M. Marley, W. B. Hubbard et al.: ApJ **491**, 856 (1997)
6. A. Burrows, M. S. Marley, C. M. Sharp: ApJ **531**, 438 (2000)
7. G. Chabrier, I. Baraffe, F. Allard, P. H. Hauschildt: ApJ **542**, 464 (2000)
8. X. Delfosse, C. G. Tinney, T. Forveille et al.: A&A **327**, L25 (1997)
9. T. R. Geballe, S. R. Kulkarni, C. E. Woodward, G. C. Sloan: ApJ **467**, L101 (1996)
10. D. A. Golimowski, C. J. Burrows, S. R. Kulkarni et al.: AJ **115**, 2579 (1998)
11. N. Jacquinet-Husson, E. Arié, J. Ballard et al.: JQSRT **62**, 205 (1999)
12. H. R. A. Jones, A. J. Longmore, R. F. Jameson, C. M. Mountain: MNRAS **267**, 413 (1994)
13. H. R. A. Jones, T. Tsuji: ApJ **480**, L39 (1997)
14. J. D. Kirkpatrick, F. Allard, T. Bida et al.: ApJ **519**, 802 (1999)
15. J. D. Kirkpatrick, C. A. Beichman, M. F. Skrutskie: ApJ **476**, 311 (1997)
16. J. D. Kirkpatrick, I. N. Reid, J. Liebert et al.: ApJ **519**, 834 (1999).
17. J. D. Kirkpatrick, I. N. Reid, J. Liebert, et al.: AJ **120**, 447 (2000)
18. S. K. Leggett: ApJS **82**, 351(1992)
19. S. K. Leggett, T. R. Geballe, X. Fan et al.: ApJ **536**, L35 (2000)
20. S. K. Leggett, D. W. Toomey, T. R. Geballe, R. H. Brown: ApJ **517**, L139 (1999)
21. J. Liebert, I. N. Reid, A. Burrows et al.: ApJ **533**, L155 (2000)
22. K. Matthews, T. Nakajima, S. R. Kulkarni, B. R. Oppenheimer: AJ **112**, 1678 (1996)
23. T. Nakajima, B. R. Oppenheimer, S. R. Kulkarni et al.: Nature **378**, 463 (1995)
24. T. Nakajima, T. Tsuji, T. Maihara et al.: PASJ **52**, 87 (2000)
25. B. R. Oppenheimer, S. R. Kulkarni, K. Matthews, M. H. van Kerkwijk: ApJ **502**, 932 (1998)
26. I. N. Reid, J. D. Kirkpatrick, J. Liebert et al.: ApJ **521**, 613 (1999)
27. L. S. Rothman: HITEMP CD-ROM (ONTAR, Andover 1997)
28. M. A. Strauss, X. Fan, J. E. Gunn et al.: ApJ **522**, L61 (1999)
29. C. G. Tinney: Nature **397**, 37 (1999)
30. T. Tsuji: In: *Very Low-Mass Stars and Brown Dwarfs in Stellar Clusters and Associations* ed. R. Rebolo (Cambridge Univ. Press, Cambridge 2000) in press
31. T. Tsuji, K. Ohnaka: In: *Elementary Processes in Dense Plasmas* eds. S. Ichimaru & S. Ogata (Addison-Wiley, Reading 1995) pp.193–200
32. T. Tsuji, K. Ohnaka, W. Aoki: A&A **305**, L1, (1996)
33. T. Tsuji, K. Ohnaka, W. Aoki, T. Nakajima: A&A **308**, L29 (1996)
34. T. Tsuji, K. Ohnaka, W. Aoki: ApJ **520**, L119 (1999)
35. Z. I. Tsvetanov, D. A. Golimowski, W. Zheng et al.: ApJ **531**, L61 (2000)

Alkali Metals and the Colour of Brown Dwarfs

A. Burrows

Department of Astronomy and Steward Observatory,
The University of Arizona, Tucson, AZ 85721, USA

Abstract. I summarize some of the consequences for the optical and very-near-infrared spectra of T dwarfs (in particular) and brown dwarfs (in general) of their possible dominance by the neutral alkali metal lines. As a byproduct of this study, I estimate the true optical colour of "brown" dwarfs.

1 Introduction

The early discovery phase for L dwarfs and T dwarfs has ended and a major focus is now on their characterization. The atmospheres of brown dwarfs are dominated by H_2, H_2O, CH_4, NH_3, the neutral alkali metals, and grains, but how theory translates this basic knowledge into effective temperatures, gravities, and compositions has yet to be determined. Establishing the spectral and colour diagnostics that are most appropriate for L/T studies is complicated by ambiguities in the cloud/grain models and a paucity of opacity data. In particular, though T dwarfs are being informally defined by their methane features at 1.7 μm and 2.2 μm, the methane database itself is far from complete. The methane opacities on the red side of the H band are certainly in error by a factor of 3 to 5 (witness Gliese 229B[1]) and the hot bands are completely missing. The latter means that even the sign of the opacity's dependence upon temperature may be in error. Nevertheless, there has been great overall progress towards understanding what makes these objects unique and what their spectra are telling us. In this paper, I sidestep a comprehensive study of these issues and summarize three interesting topics in brown dwarf theory that have emerged of late. They are 1) what determines T dwarf spectra shortward of 1.0 micron, 2) what is the true colour of a "brown" dwarf, and 3) what is the effect of heavy element depletion ("rainout") on the abundance profiles of the neutral alkali metal atoms. A subtext of this contribution is the central importance of the alkali metals in spectrum formation.

2 The Short-Wavelength Spectra of T Dwarfs

Employing the scheme of Burrows, Marley, and Sharp[2] (hereafter BMS), we can derive the neutral alkali opacities as a function of wavelength. Figure 1 depicts the abundance-weighted opacities of the dominant neutral alkali metal lines at 1500 K and 1 bar. This opacity spectrum has a bearing on the suggestion by BMS that the strong continuum absorption seen in all T dwarf spectra in

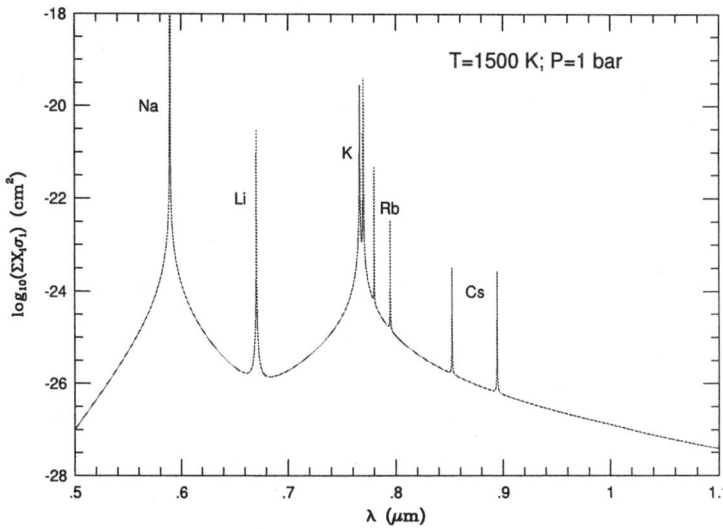

Fig. 1. Plotted is the abundance-weighted cross section spectrum for the neutral alkali metals Na, K, Cs, Rb, and Li at 1500 K and 1 bar pressure, using the theory of BMS. The most important spectral lines for each species are clearly marked.

the near-infrared from 0.8 μm to 1.0 μm , previously interpreted as due to an anomalous population of red grains[3] or in part due to high-altitude silicate clouds[4], is most probably due to the strong red wings of the K I doublet at ∼7700 Å. This is demonstrated in Fig,. 2, in which several possible theoretical spectra are compared with the observed spectrum for Gliese 229B in the near-infrared[2]. Tsuji et al. [5] also identified the K I doublet as one of the agents of absorption shortward of one micron, but they needed silicate grains as well to reproduce the Gliese 229B observations. BMS conclude that the K I resonance doublet alone is responsible, though, given the remaining ambiguity in its line shape, one can not completely eliminate the presence of grains as secondary agents.

As Fig. 2 suggests, the BMS theory also explains the WFPC2 I band (M_I ∼20.76; theory = 21.0) and R band (M_R ∼ 24.0; theory = 23.6) measurements made of Gl 229B[6], with the Na D lines at 5890 Å helping to determine the strength of the R band. BMS predicted not only that there would be a large trough in a T dwarf spectrum at 7700 Å due to the K I resonance, but that the spectrum of a T dwarf would peak between the Na D and K I absorption troughs at 5890 Å and 7700 Å, respectively. This prediction was recently verified by Liebert et al. [7] for the T dwarf SDSS 1624+00.

Furthermore, the 1.17 μm and 1.24 μm subordinate lines of excited K I have been identified in T dwarfs[8–10]. Since these subordinate lines are on the crown of the J band, they allow one to probe the deeper layers at higher temperatures. Figure 3 portrays for a representative Gl 229B model the dependence on wave-

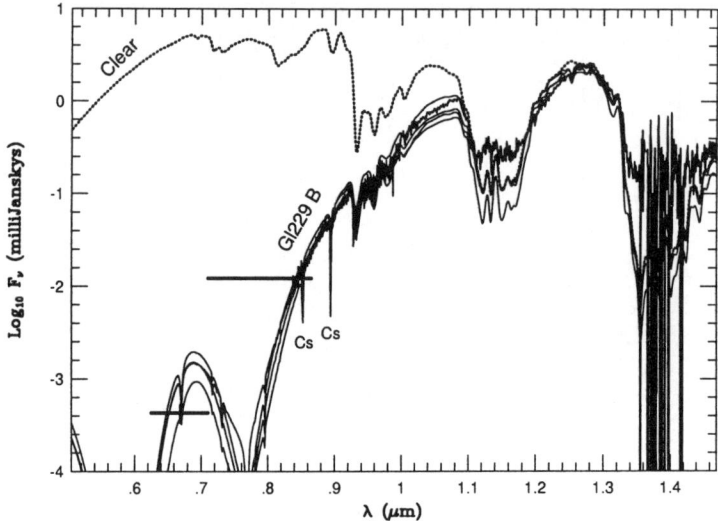

Fig. 2. The log of the absolute flux (F_ν) in milliJanskys versus wavelength (λ) in microns from 0.5 μm to 1.45 μm for Gliese 229 B, according to Leggett et al. [1] (heavy solid), and for four theoretical models (light solid) described in BMS. Also included is a model, denoted "Clear" (dotted), without alkali metals and without any ad hoc absorber due to grains or haze. The horizontal bars near 0.7 μm and 0.8 μm denote the WFPC2 R and I band measurements of Golimowski et al. [6]. Figure taken from BMS.

length of the "brightness" temperature, here defined as the temperature at which the photon optical depth is 2/3. Such plots clearly reveal the temperature layers probed with spectra and provide a means to qualitatively gauge composition profiles. Specifically, for the Gl 229B model, the detection of the subordinate lines of potassium indicates that we are there probing to ~1600 K, while the detection of the fundamental methane band at 3.3 μm (not shown in Fig. 3) means that we are probing to only ~600 K.

3 The Colour of Brown Dwarfs

Figure 1 shows that the Na D doublet should dominate the optical portion of the spectrum. Since it suppresses the green wavelengths and "brown" is two parts red, one part green, and very little blue, brown dwarfs should not be brown. In fact, our recent calculations suggest that they are red to purple, depending upon the exact shape of the line wings of Na D, the abundance of the alkalis, the presence of high-altitude clouds, and the role of water clouds at lower $T_{\rm eff}$s ($\lesssim$ 500 K). A mixture of red and the complementary colour to the yellow of the Na D line makes physical sense. It is the *complementary* colour, not the *colour*, of the Na D line(s) because Na D is seen in absorption, not emission. Indeed, the recent measurement of the spectrum of the L5 dwarf 2MASSW J1507 from 0.4

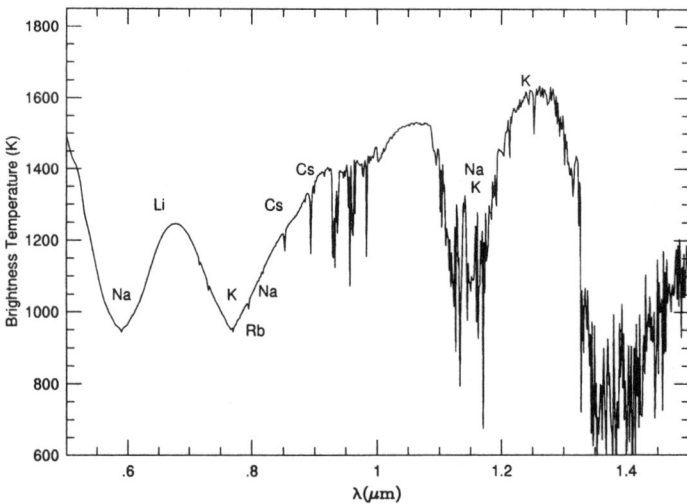

Fig. 3. The brightness temperature in Kelvin versus wavelength in microns from 0.5 μm to 1.5 μm of a simple model of Gliese 229B. The brightness temperature for a given wavelength is defined as the temperature of the layer at which $\tau_\lambda = 2/3$. The identity of the alkali metal atom responsible for a given feature is indicated. See text for discussion.

μm to 1.0 μm (I.N. Reid and J.D. Kirkpatrick, in preparation) indicates that this L dwarf is magenta in (optical) colour. This is easily shown with a program that generates the RGB equivalent of a given optical spectrum (in this instance, R:G:B::1.0:0.3:0.42, depending upon the video "gamma"). Hence, after a quarter century of speculation and ignorance, we now have a handle on the true colour of a brown dwarf — and it is not brown.

4 Rainout and the Alkali Metals

As shown by Burrows and Sharp[11], Fegley and Lodders[12], and Lodders[13], the alkali metals are less refractory than Ti, V, Ca, Si, Al, Fe, and Mg and survive in abundance as neutral atoms in substellar atmospheres to temperatures of 1000 K to 1500 K. This is below the 1600 K to 2500 K temperature range in which the silicates, iron, the titanates, corundum, and spinel, etc. condense and rainout. The rainout of refractory elements such as silicon and aluminum ensures that Na and K are not sequestered in the feldspars high albite ($NaAlSi_3O_8$) and sanadine ($KAlSi_3O_8$) at temperatures at and below 1400 K, but are in their elemental form down to ~1000 K. Hence, in the depleted atmospheres of the cool T dwarfs and late L dwarfs, alkali metals quite naturally come into their own. Figures 4 and 5 demonstrate the role of rainout by depicting the profiles of the relative abundances of the main reservoirs of the alkali metals, with and without rainout as crudely defined in reference [11]. As is clear from a comparison of these

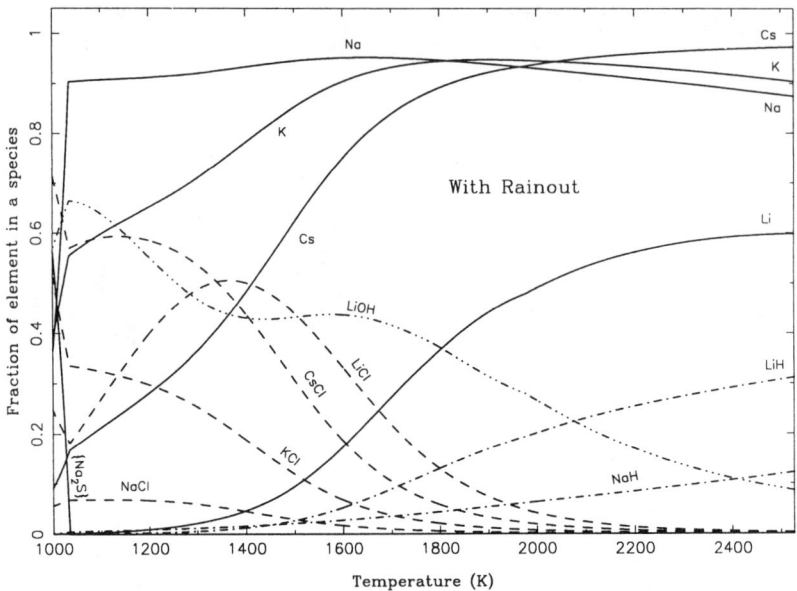

Fig. 4. The fractional abundances of different chemical species involving the alkali elements Li, Na, K and Cs for a Gliese 229B model, with rainout as described in Burrows and Sharp[11]. The temperature/pressure profile for a T_{eff} =950 K and $g = 10^5$ cm s^{-2} model, taken from Burrows et al. [17], was used. Each curve shows the fraction of the alkali element in the indicated form out of all species containing that element. All species are in the gas phase except for the condensates, which are in braces { and }. The solid curves indicate the monatomic gaseous species Li, Na, K and Cs, the dashed curves indicate the chlorides, the dot-dashed curves indicate the hydrides and the triple dot-dashed curve indicates LiOH. Due to rainout, at lower temperatures there is a dramatic difference with the no–rainout, complete equilibrium calculation (Figure 5); high albite and sanidine do not appear, but instead at a much lower temperature the condensate Na$_2$S (disodium monosulfide) forms, as indicated by the solid line in the lower left of the figure. The potassium equivalent, K$_2$S, also forms, but it does so below 1000 K and is not indicated here. The difference between this Figure and Figure 5 is that almost all the silicon and aluminum have been rained out at higher temperatures, so that no high albite and sanidine form at lower temperatures. Figure taken from BMS.

two figures, rainout and depletion of heavy metals may result in a significant enhancement in the abundances at altitude (lower temperatures) of the neutral alkali metal atoms, in particular sodium and potassium.

Figure 2 demonstrates the naturalness with which the potassium resonance lines alone fit the observed near-infrared/optical spectrum of Gl 229B. Curiously, in the metal-depleted atmospheres of T dwarfs, the reach of the K I doublet is one of the broadest in astrophysics, its far wings easily extending more than 1500 Å to the red and blue. With rainout, below ~1000 K both sodium and potassium exist as sulfides (Na$_2$S and K$_2$S)[13]. Without rainout, complete chemical equilibrium

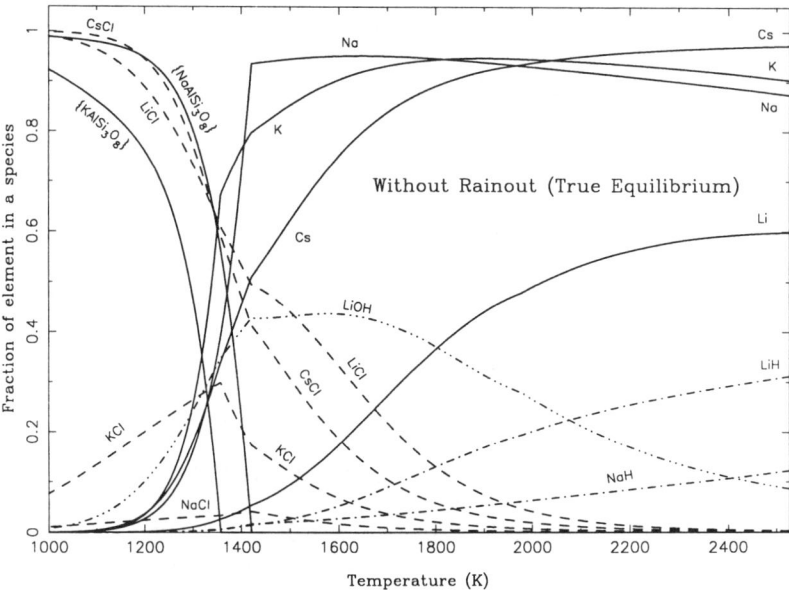

Fig. 5. The fractional abundances of different chemical species involving the alkali elements Li, Na, K and Cs for a Gliese 229B model, assuming complete (true) chemical equlibrium and no rainout (disfavored). The temperature/pressure profile for a $T_{\text{eff}} = 950$ K and $g = 10^5$ cm s^{-2} model, taken from Burrows et al. [17], was used. Each curve shows the fraction of the alkali element in the indicated form out of all species containing that element, e.g., in the case of sodium, the curves labeled as Na, NaCl, NaH and NaAlSi$_3$O$_8$ are the fractions of that element in the form of the monatomic gas and three of its compounds. All species are in the gas phase except for the condensates, which are in braces { and }. The solid curves indicate the monatomic gaseous species Li, Na, K and Cs and the two condensates NaAlSi$_3$O$_8$ and KAlSi$_3$O$_8$, i.e., high albite and sanidine, respectively, the dashed curves indicate the chlorides, the dot-dashed curves indicate the hydrides and the triple dot-dashed curve indicates LiOH. Figure taken from BMS.

at low temperatures, requires that sodium and potassium reside in the feldspars. If such compounds formed and persisted at altitude, then the nascent alkali metals would be less visible, particularly in T dwarfs. By modeling spectra with and without the rainout of the refractories and comparing to the emerging library of T dwarf spectra[10,14–16], the degree of rainout and the alkali composition profiles in brown dwarf atmospheres may be approximately ascertained.

5 Conclusion

L and T dwarf spectra are unique among "stars" and require new databases, approaches, and thinking to understand fully. Exploring as we are new worlds, we will require new tools and instincts with which to navigate. Along with accurate cloud models, methane, and water, the alkali metals hold the key to unraveling the mysteries of the substellar objects that we now know inhabit the solar neighborhood in abundance.

Acknowledgments

I thank my long-time collaborators, Jonathan Lunine, Bill Hubbard and Mark Marley for simulating input and both Davy Kirkpatrick and Neill Reid for an advanced glimpse at their stunning 2MASSW J1507 spectrum. This work was supported in part by NASA under grants NAG5-7073 and NAG5-7499.

References

1. S. Leggett, D.W. Toomey, T. Geballe, R.H. Brown: Astrophys. J. **517**, L139 (1999)
2. A. Burrows, M.S. Marley, C.M. Sharp: Astrophys. J. **531**, 438 (2000).
3. C.A. Griffith, R.V. Yelle, M.S. Marley: Science **282**, 2063 (1998)
4. F. Allard, P.H. Hauschildt, D.R. Alexander, S. Starrfield: Annu. Rev. Astron. Astrophys. **35**, 137 (1997)
5. T. Tsuji, K. Ohnaka, W. Aoki: Astrophys. J. **520**, L119 (1999)
6. D.A. Golimowski, et al. : Astron. J. **115**, 2579 (1998)
7. J. Liebert, I.N. Reid, A. Burrows, A.J. Burgasser, J.D. Kirkpatrick, J.E. Gizis: Astrophys. J. **533**, 155 (2000)
8. M.A. Strauss, et al. : Astrophys. J. **522**, L61 (1999)
9. Z.I. Tsvetanov, et al. : Astrophys. J. **531**, L61 (2000)
10. I.S. McLean, et al. : Astrophys. J. **533**, L45 (2000)
11. A. Burrows, C.M. Sharp: Astrophys. J. **512**, 843 (1999)
12. B. Fegley, K. Lodders: Astrophys. J. **472**, L37 (1996)
13. K. Lodders: Astrophys. J. **519**, 793 (1999)
14. A. Burgasser, et al. : Astrophys. J. **522**, L65 (1999)
15. A. Burgasser, et al. : Astrophys. J. **531**, L57 (2000)
16. S. Leggett, *et al.*: Astrophys. J. **536**, L35 (2000)
17. A. Burrows, M. Marley, W.B. Hubbard, J.I. Lunine, T. Guillot, D. Saumon, R. Freedman, D. Sudarsky, C. Sharp: Astrophys. J. **491**, 856 (1997)

Formation of the Optical Spectra of L Dwarfs

Y. Pavlenko

Main Astronomical Observatory of NAS, Golosiiv woods, 03680, Kyiv-127, Ukraine

Abstract. Formation processes of spectra of low-mass stars and substellar objects (L dwarfs) with $1200 < T_{eff} < 2200$ K are discussed. We show that the overall shapes of their optical spectra are governed by the strong resonance lines of K I and Na I. Mechanisms of their broadening in stellar atmospheres of late spectral classes are analyzed. The computed spectral energy distributions (SED's) of L dwarfs are compared to the observed spectra. Perspectives of Lithium and Deuterium tests to assess the new populations of the Galaxy have been considered.

1 Introduction

Recent observations have provided spectra of a new class of substellar objects ($T_{eff} < 2400$ K) which differ from spectra of M dwarfs (Fig. 1). The VO and TiO bands in the optical spectra are strongly weakened or even disappear. Undoubtedly in the near future, these dwarfs will be accepted as belonging to a new spectral class.

For the time being at least two systems of spectral classification have been proposed for L dwarfs ([1–3]). Probably, BRI0021-0214 ($T_{eff} \approx 2200$ K) lies at a boundary line between M- and L dwarfs (see Fig. 2 in [6]).

2 Dwarfs of Spectral Classes M, L, T

In the past few years we have understood that "the bottom of the Main Sequence" is populated by different objects:

M dwarfs are objects (stars + young brown dwarfs) with $4000 > T_{\text{eff}} > 2200$ K. Their spectra are governed by molecular bands of TiO and VO (in the optical part of the spectrum) , H_2O (in the red) [7–9].

L dwarfs are objects (brown dwarfs + stars) with $1200 < T_{\text{eff}} < 2200$ K (cf. [2]). Optical spectra of L dwarfs are governed by the K I + Na I lines [10] and molecular bands of CrH + FeH (in the optical spectrum), CO and H_2O (in the infrared) ([11,12]). GD165B seems to be the first L dwarf discovered approx. 10 years ago ([13]).

T dwarfs have planet-like near-infrared spectra with $T_{\text{eff}} < 1200$ K. Their spectra are governed by the "dusty opacities" + methane and H_2O bands in absorption ([14,15]).

Planets are objects with masses $M < 13$ M_j which preserve their initial deuterium.

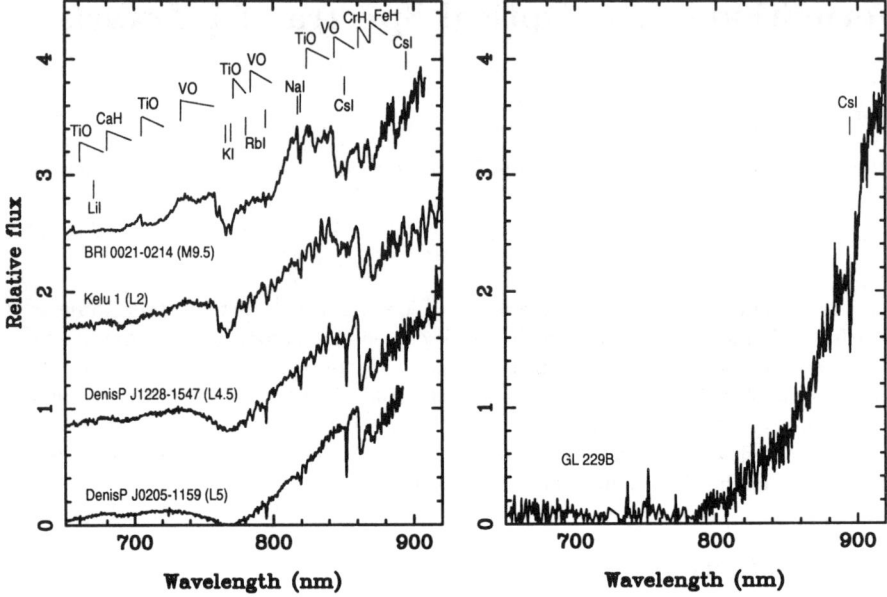

Fig. 1. Far-red optical spectra of some M, L and T dwarfs are shown in the left and right panels. The observations are those of Kelu 1, Denis-P J1228–1547 and Denis-P J0205–1159 (KeckII spectrum). Data for BRI 0021-0214 and Gl 229B have been collected from Martín et al. [6] and from Schultz et al. [4] respectively. Spectral types for the L dwarfs are given following the classification of Martín et al. [2]. Spectra of L dwarfs in the left panel have been shifted by 0.8 units for clarity. Identification of some atomic and molecular features is provided in the top (from [5]).

3 Procedure

Our computations of synthetical spectra of L dwarfs are carried out by program WITA5, which is a version of the program WITA31 [9]. The principal modifications are aimed to incorporate "dusty effects" that may affect the chemical equilibrium and radiative transfer processes in very cool atmospheres.

We use a grid of the "dusty" (B- and C-types) LTE model atmospheres of Tsuji [16]. These models were computed for the case of the "dust-gas" segregation phase. Other details of computations are given in [11,5].

In L dwarf atmospheres the additional opacity (AdO) maybe due to molecular and/or dust absorption and/or scattering. We have modeled the additional opacity with a simple law of the form $a_\circ \, (\nu/\nu_\circ)^N$, with $N = 1$ - 4 (see [5] for more details).

4 Results

4.1 Chemical Equilibrium: Atoms, Molecules, Dust

Cool plasma of L dwarfs atmospheres consists of the mix of free electrons, neutral atoms, ions, molecules and dust particles. Alkali metals are the main donors of free electrons there because they have low ionization potentials. In general, LTE densities of the free electrons in the outermost layers of L dwarfs atmospheres are low enough($n_e \leq 1 - 10^4$ cm^{-3}). Only in the photospheric layers n_e increases up to the "usual" values $> 10^{10}$ cm^{-3}. The NLTE effects of *overionization* of alkali metals [17,18] cannot be efficient in very cool atmospheres because the atoms of neutral species are in ground state mainly. Photoionization processes from low lying levels of metals are efficiently blocked there due to the absorption/scatter of UV photons by dust particles and/or molecules.

Processes of formation of molecules are of importance in atmospheres of the coolest dwarfs. Naturally, the formation of molecules densities of neutral atoms are reduced ([19]). In fact the molecules are main reservoir of "the sink" of the atoms of some metals in M dwarfs atmospheres.

It is worth to note that alkali metals show different abilities to form molecular species ([20]). In general, atoms of Na and K are not chemically active enough to be very depleted even in the L dwarf atmospheres with $T_{eff} = 1200$ K (Fig. 2). On the other hand, Li atoms may be depleted into molecular species even in M dwarf atmospheres ([19]). In the outer part of L dwarfs atmospheres Li atoms are bound into LiCl, LiBr and LiOH molecules (Fig. 3).

Processes of formation of molecular species may affect the densities of the atomic hydrogen – the most abundant molecule in the atmospheres of the comparatively warmer M dwarfs. Computations of the chemical equilibrium show that in atmospheres of L dwarfs hydrogen atoms exist mainly in the form of different molecular species (H$_2$, HCN, H$_2$O, etc.). In that case $n(H^+) << n(H) < n(H_2)$.

In high pressure and cool layers of photospheres of L dwarfs, some atoms are depleted also in grain particles [21]. Indeed, the chemical equilibrium computation shows that for some species a regime of *oversaturation* may occur, i.e., molecular densities of these species should be reduced due to their depletion into grain particles via "dust-gas" phase transition. A comparison of the "gas-dust equilibrium" and "chemically equilibrium" molecular densities computed for some B-model atmospheres of T. Tsuji [21] is shown in the Fig. 4. Partial pressures of saturation for some species were taken from Gurvitz et al. [22]. Naturally, even in the frame of our simplest approach, the molecular densities of Ti, TiO (as well VO, see Pavlenko 1998) should be reduced there in comparison with the pure chemical equilibrium computations. Jones & Tsuji [23] discussed an impact of those depletion processes on spectra of the latest M dwarfs.

4.2 Spectra of L dwarfs

Nevertheless, due to depletion processes of Ti and VO atoms into grain particles, the optical spectra of L dwarfs becomes simpler than that of M dwarfs. Their

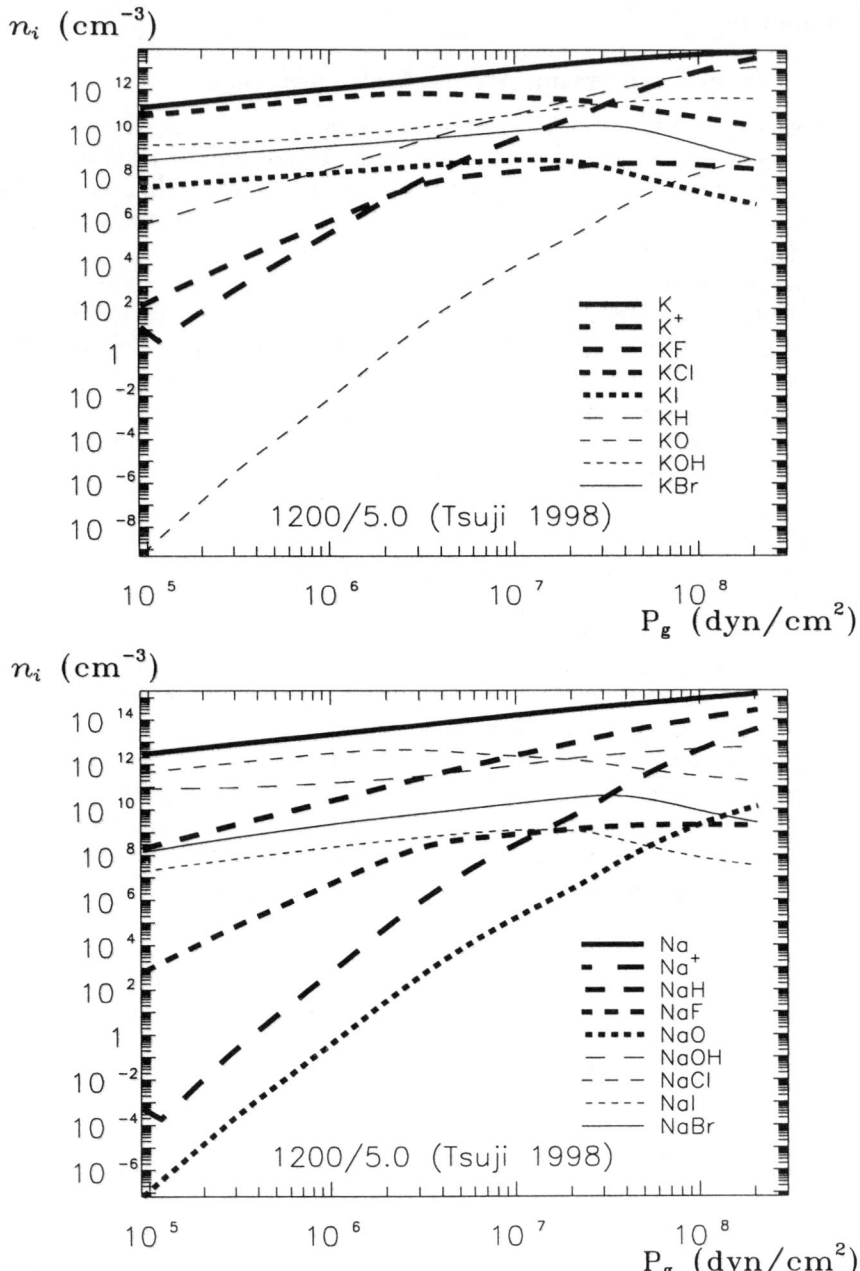

Fig. 2. Molecular densities of K and Na contained species in 1200/5.0 C-model atmosphere of T.Tsuji [16].

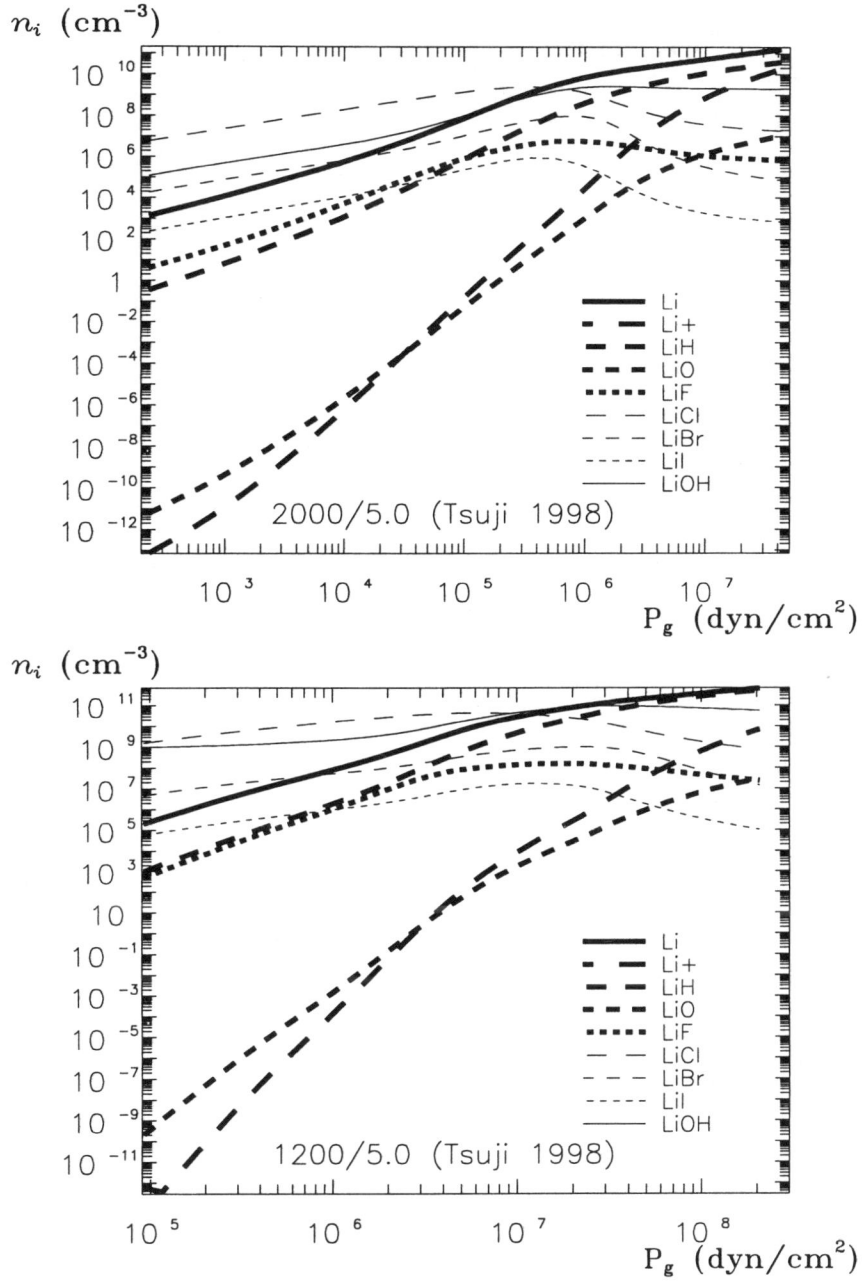

Fig. 3. Molecular densities of the Li contained species in 2000/5.0 and 1200/5.0 C-model atmospheres of T.Tsuji [16].

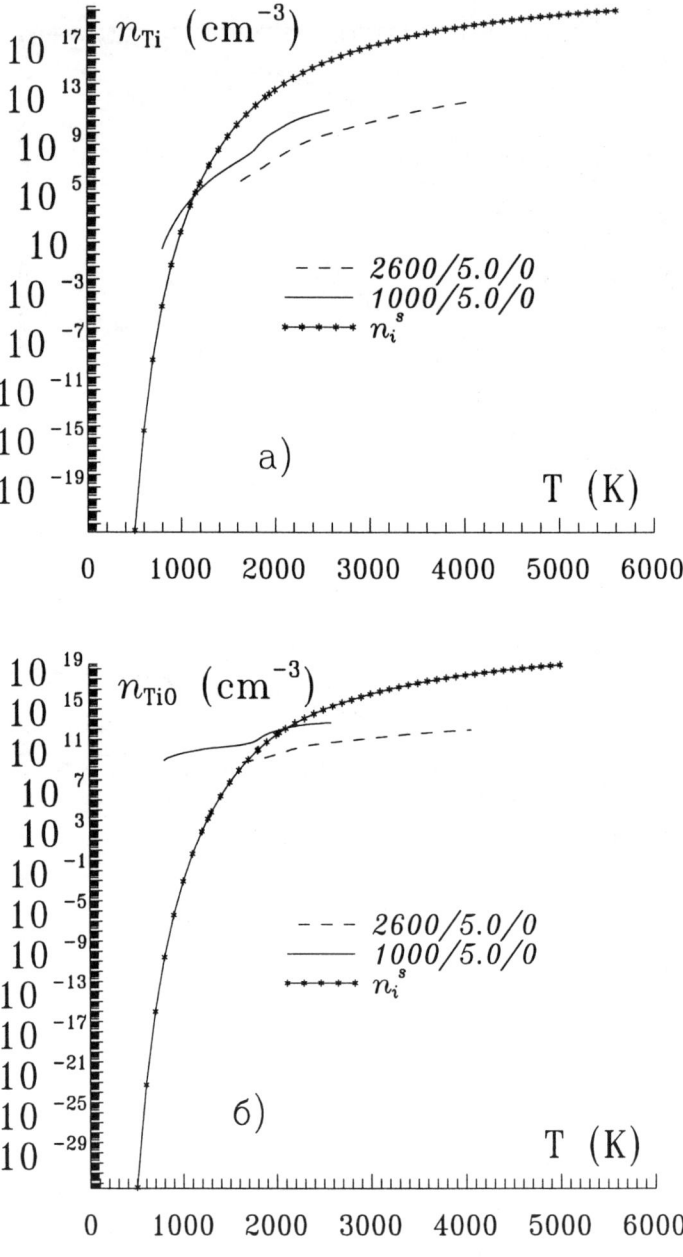

Fig. 4. Comparison of molecular densities of Ti and TiO computed for the case of the chemical equilibrium and the "dust-gas" phase transition.

overall spectral energy distributions (SED) are governed by absorption of resonance doublets of K I and Na I which have their pressure-broadened wings [10] (see computed damping constants in Fig. 5) extended up to thousands Å(Fig. 6). Their formally-computed equivalent widths may be of several thousand of Å[24].

The broadening of the resonance lines depends on the collisional interactions of the atoms with the most abundant species, i.e. neutral H and He atoms, and H_2 molecules. We suggest that (at least in the first approach) the profiles of the strong absorption lines may be fitted by Voigt function $H(a,v)$, as usually, $a = \gamma/\Delta\nu_D, v = \Delta\nu/\Delta\nu_D$ and $\Delta\nu_D$ - doppler width of lines.

In first versions of the WITA program the approximate formulas of Kurucz [25] were used to compute damping constants γ:

$$\gamma = E * n(H) * \gamma_6(H) * (1 + 0.45 * n(He)/n(H) + 0.83 * n(H_2)/n(H)), \quad (1)$$

where $n(H), n(He)$, and $n(H_2)$ are molecular densities of H, He and H_2, respectively; $\gamma_6(H)$ and E are constant of van der Waals broadening due to the collisional interactions with hydrogen atoms and "correction factor" from the Unsold's [26] formula. In that way we used the Unsold's approximation for all kinds of the van der Waals broadening. For alkali lines in solar-like atmospheres Andretta et al. [27] computed $E \approx 1.5$. However, the value 1.5 corresponds to the interaction of alkali atoms with the neutral hydrogen atoms. In the case of L dwarfs atmospheres $n(H) < n(H_2)$, we suggest that $E \approx 1$ also for the "alkali atom—molecular hydrogen" interactions.

Pavlenko et al. [19] from the fit of the profiles of the subordinate Na I lines $\lambda\lambda$ 819.6986, 819.7016, 818.5443 nm to the observed spectra of the late M dwarfs found that the formula (1) overestimates the contributions of molecular hydrogen into the broadening of the absorption lines. Better results were obtained with formula

$$\gamma = E * n(H) * \gamma_6(H) * (1 + 0.42 * n(He)/n(H)), \quad (2)$$

i.e. without account of broadening of atomic lines due to the interaction with molecules H_2.

Note, that the damping constants computed with formulas (1) and (2) differ substantially – up to factor 100 in the case of 1200/5.0 model atmospheres (Fig. 5). Then, in atmospheres of L dwarfs $0.42*n(He)/n(H) >> 1$, i.e line broadening processes due to interactions with neutral He atoms dominate.

In general, profiles of resonance doublets of K and Na depend on a few parameters, i.e. $\kappa_l = f(T_{\text{eff}}, \log g, \mu, ...)$ [5]. Furthermore, Pavlenko et al. [5] show that optical spectra of L dwarfs are affected by the additional ("dusty") absorption and/or scattering (Fig. 7). Fortunately, the asymetrical shapes of the SEDs provide strong enough constraints for the models of the "dusty" properties of L dwarf atmospheres (see [5]).

We used the simplest approach to fit computed spectra to the observed profiles of K I and Na I lines. Further attempts should be done in the frame of more

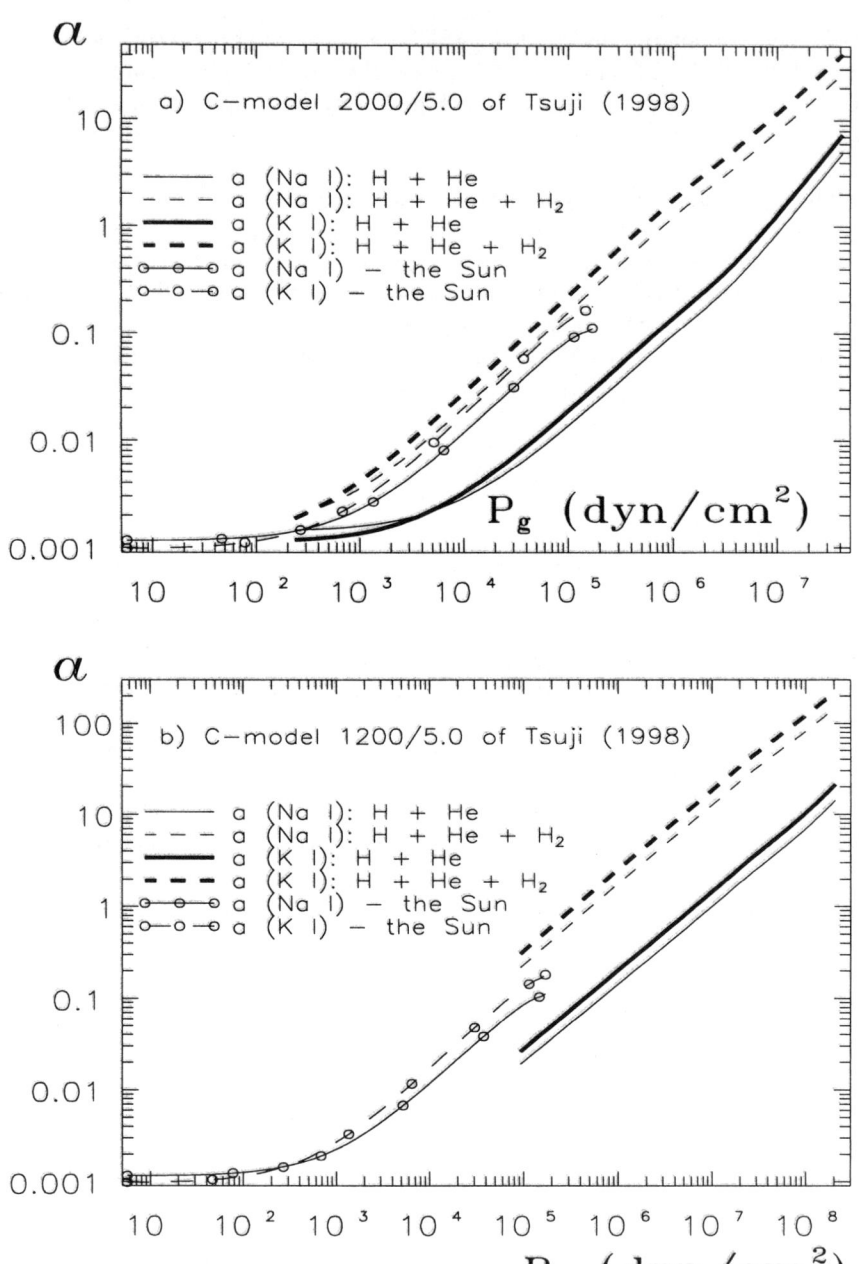

Fig. 5. Damping constants of K I and Na I resonance doublets computed for the model atmosphere of the Sun and 2000/5.0 and 1200/5.0 C-model atmospheres.

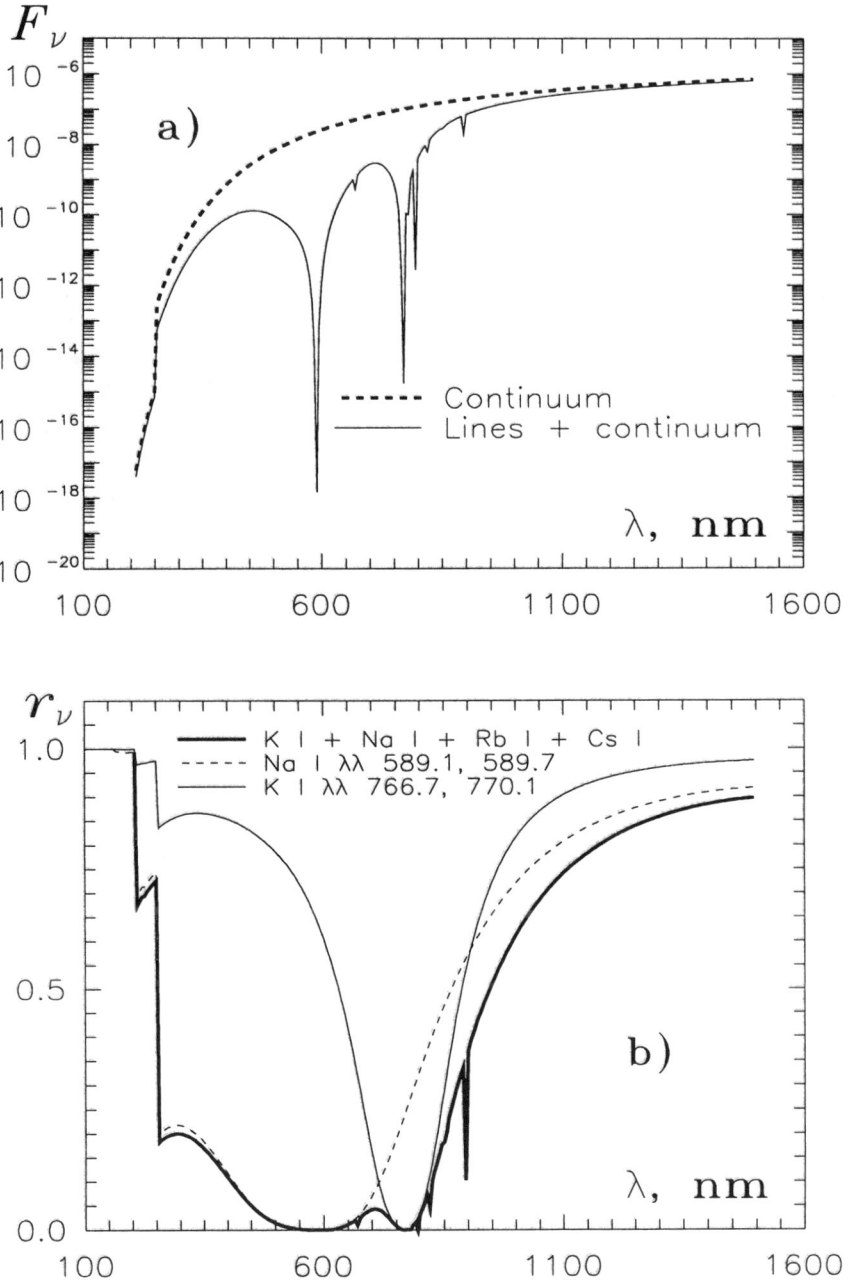

Fig. 6. Fluxes F_ν^l and residual fluxes $r_\nu = F_\nu^l/F_\nu^c$ of K I and Na I resonance doublets computed for the model 1200/5.0 C-model atmosphere of Tsuji (1998).

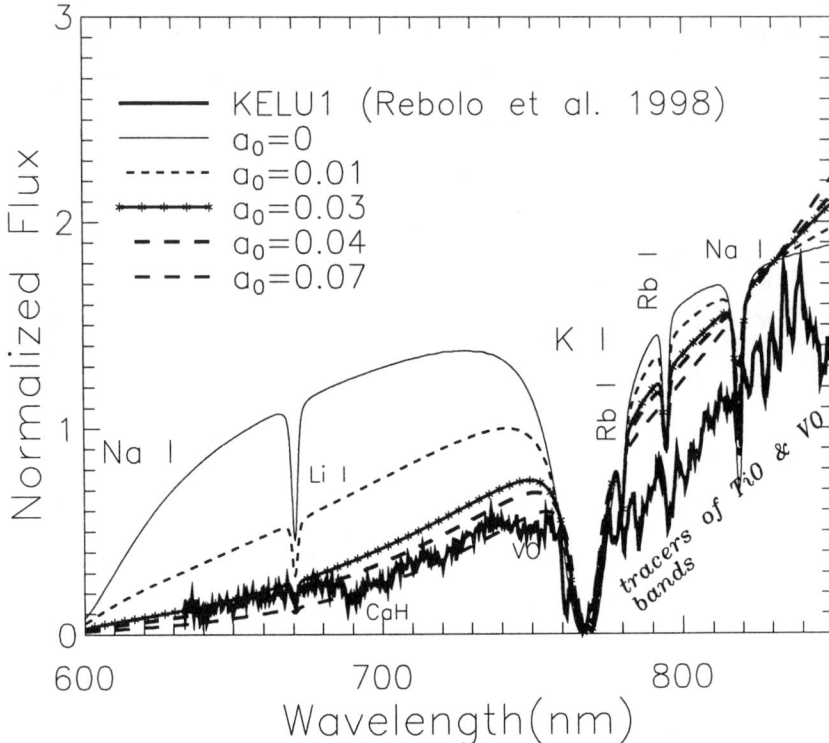

Fig. 7. The impact of the additional ("dusty") opacity on the spectral energy distribution of Kelu1 ($T_{eff} \approx 2000$ K). Opacities due to the absorption by molecular band are not taken into account here. Note – the overall shape of SEDs are governed by Na I + K I absorption; the intensities of alkali lines are severely affected by the "dusty" opacity. Some weak tracers of CaH, VO and TiO band systems may be recognized in Kelu1 spectrum (see [5] for more details).

sophisticated theory of the resonance lines formation. Recently, Nefedov et al. [28] (see also [29]) show that the broadening of the resonance lines due to the interaction with the H_2 molecule should be taken into account (see subsection 4.4).

4.3 Algorithm of L dwarf spectra formation

Spectra of the cool L dwarf spectra provide the strong enough CrH bands of the $A^6\Sigma^{(+)}$-$X^6\Sigma^{(+)}$ system [30] and Fe H bands [12]. If we do not include into computations the absorption of K I and Na I resonance lines then CrH and FeH bands dominate in theoretical spectra of L dwarfs (Fig. 8, top). However, the SED's computed accounting for K I and Na I resonance lines are very different from those. Huge absorption in the resonance lines governs the spectra shape over the wide spectral regions. If we include both molecular bands and resonance

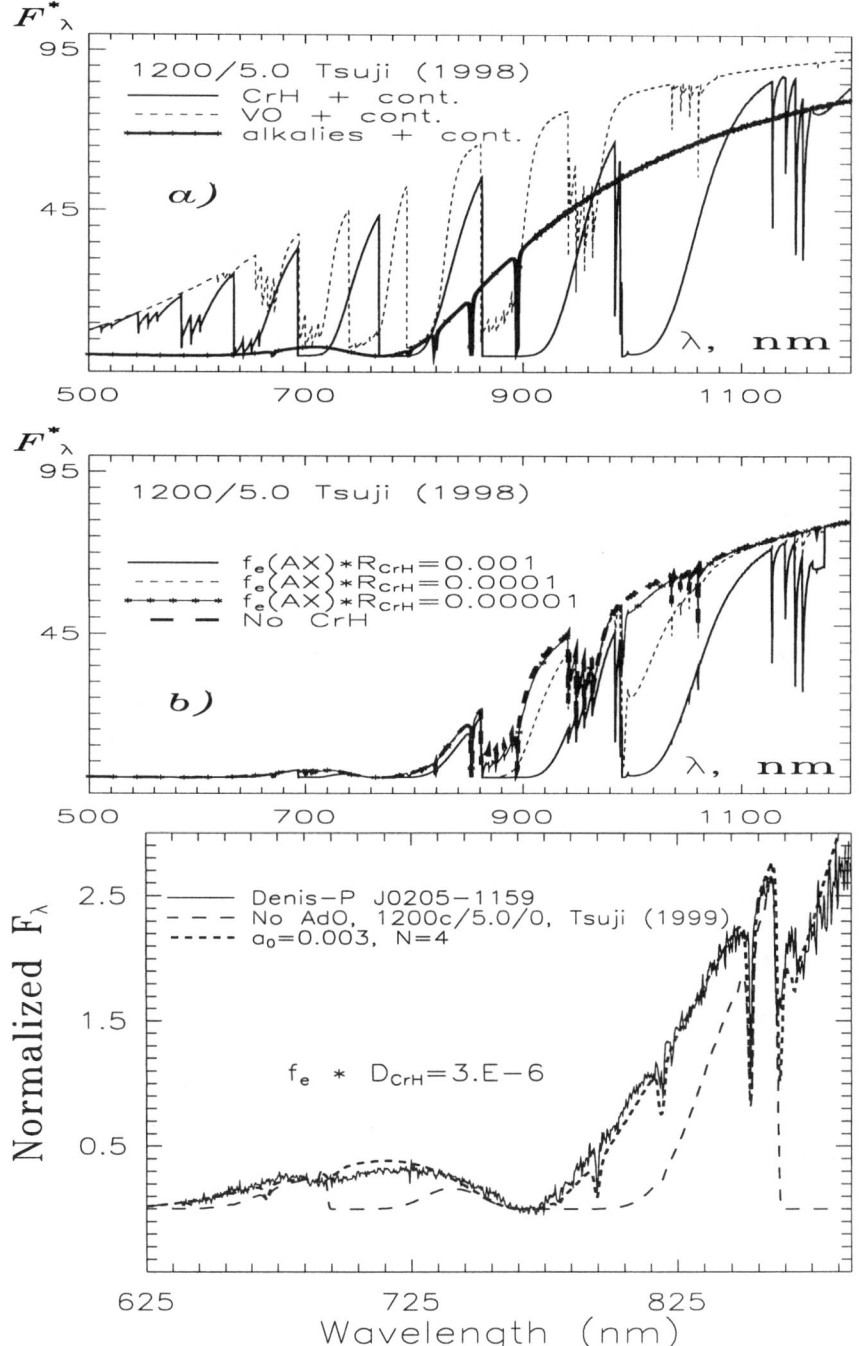

Fig. 8. Algorithm of formation of L dwarf 1200/5.0 (DenisP J0205-1159) spectrum.

Table 1. Equivalent widths of the Li I resonance doublet at 670.8 nm computed for the C-type Tsuji's [16] C-model atmospheres, cosmic Li abundance (log N(Li) = 3.2) and gravity log g = 5.0. (from [5])

T_{eff}	a_o		
	0.00	0.01	0.10
(K)	W_λ(Å)		
1000	17	8	0.6
1200	30	12	0.7
1400	42	21	0.9
1600	40	24	1.6
2000	23	16	3.6

atomic lines we obtain very reliable picture (Fig. 8, bottom). Furthermore, the additional opacity (AdO) concept allows us to improve even the fine details of L dwarf spectra.

4.4 Broadening of resonance lines and T_{eff} scale of L dwarfs

Comparison of computed and observed spectra of L dwarfs allows us to verify our procedure of the line-profiles computation. Indeed, there we should model very extended wings of the resonance lines. Our first attempts to fit observed L dwarfs spectra are described in Pavlenko et al. [5]. We obtained good fits to optical L dwarf spectra over a wide region of effective temperatures. Furthermore, we found good agreement with other authors for the case of early L dwarfs (T_{eff} ~ 2000 K for Kelu1). However, we obtained lower effective temperatures for the coolest L dwarfs (T_{eff} ~ 1200 K for Denis P J0205-1159, other authors gave 1400-1600 K [2]). Still we found, that the the T_{eff} might be higher, if we put $E = 3$-4. Namely, for Denis P J0205-1159 we obtained "the best fits" also for T_{eff} ~ 1400 K with correction factor of the van der Waals broadening $E = 3$ – 4. As noted previously, in Pavlenko et al. [5], we ignored the broadening of resonance lines by molecular hydrogen. Now we can estimate the impact of that assumption on our results:

Upper limit of molecular hydrogen densities is $n(H)/2$. Then, new computations of constants of the van der Waals broadening in the frame of non-classical theory give $\gamma(H_2) = 0.1 * \gamma^u(H_2)$, where $\gamma^u(H_2)$ is a constant of interaction computed for the Unsold's approach. In that way we obtain $0.1/2 * n(H) * \gamma^u(H_2)$. That value is of the same order as the broadening constant due to the interaction with neutral helium atoms $n(He) * \gamma^u(He) = 0.1 * n(H) * \gamma^u(He)$, because $\gamma^u(He)/\gamma^u(H) \sim 1$. Hence, in [5] we underestimated the damping constant in the atmosphere of Denis P J0205-1159 at least by factor 2, and the T_{eff} of the L dwarfs should be a bit higher, ~ 1400 K.

The effect for hotter L dwarfs should be less pronounced because in their atmospheres $n(H_2) < n(H)/2$; thus, the main agent of the resonance lines broadening there is neutral helium.

4.5 Lithium test

Some of the known brown dwarfs are actually recognized by the detection of the Li I resonance doublet in their spectra (see [31,32,3]). Indeed, temperatures in the interiors of brown dwarfs are not high enough to burn lithium. To distinguish brown dwarfs from stars among samples of M- and early L dwarfs "the lithium test" proposed by Rebolo et al. [33] can be used.

Our computations show that lithium lines formed in L dwarf atmospheres are very sensitive to the additional absorption (AdO) that we need to incorporate in the spectral synthesis if we want to model the observed broad spectral energy distributions. In Table 1, we give the predicted equivalent widths W_λ of the Li I resonance doublet at 670.8 nm for C-model atmospheres (2000–1000 K) of Tsuji [16]. We found:

– In the AdO-free case (second column in the table), we would expect for the "cosmic" values of log N(Li) rather strong neutral Li resonance lines in the spectra of objects as cool as DenisP J0205–1159 and Gl 229B.

– The chemical equilibrium of Li-contained species still allow a sufficient number of Li atoms to produce a rather strong resonance feature.

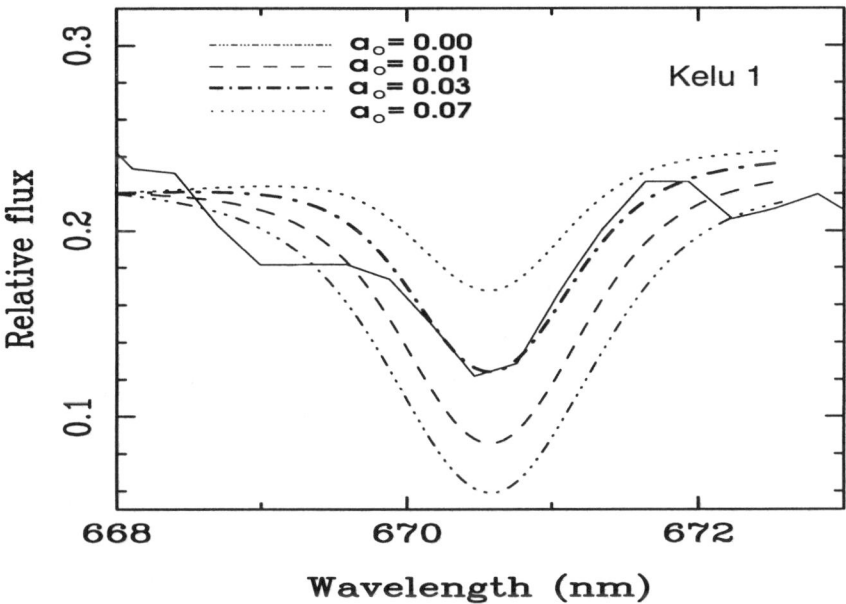

Fig. 9. The fit of the synthetical spectra to the Li I observed in Kelu1 (from [5]).

– Our computations indicate that L dwarfs with moderate dust opacities should show the Li I resonance doublet if they had preserved this element from nuclear burning and hence the lithium test can still be applied.

– Temporal variations of the dusty opacities may originate some kind of "meteorological" phenomena occurring in these cool atmospheres. Lithium lines (as well other lines) may be severely affected by the additional opacity (Fig. 9, see also Pavlenko et al. [5] for more details).

4.6 Deuterium test

The "deuterium test" was proposed recently to tell substellar objects which can burn deuterium from the planets which cannot have any nuclear burning (see discussion by Martin in this volume). Deuterium begins to burn in the interiors for $T \approx 10^6$ K. It means that objects with $M < 0.013$ $M_\odot$ should preserve their deuterium from the time of formation [35]. On the other hand, the evolutionary changes of the deuterium abundance in the atmospheres of more massive substellar objects should provide good constraints on their evolutionary status ([34]).

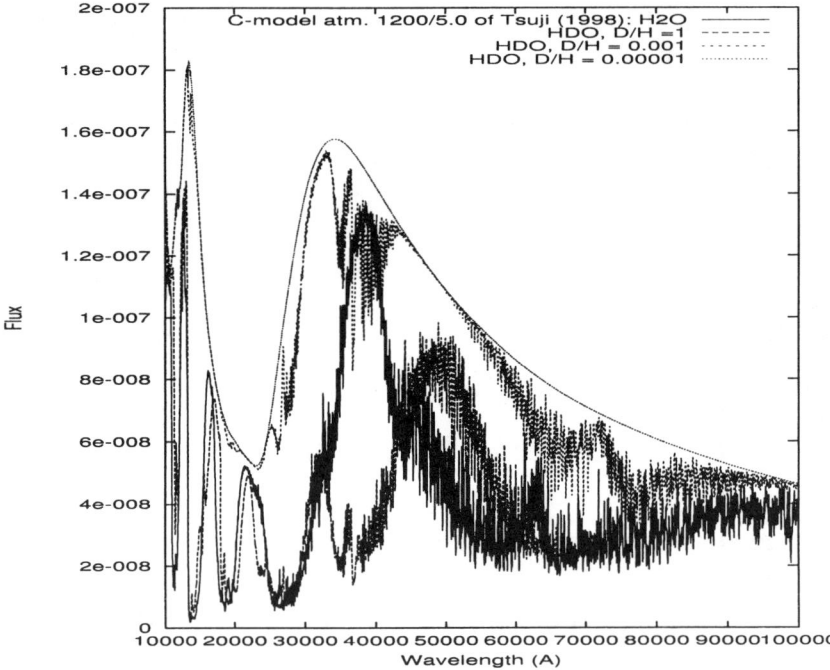

Fig. 10. Spectral energy distributions of L dwarf 1200/5.0 computed with account of absorption of H_2O, HDO. Computations of HDO spectra are carried out for D/H = 1, 0.01 and 0.00001 ("cosmic value").

Water wapour (H_2O) densities are very high in L dwarf atmospheres. Furthermore, water absorption provides very strong bands in the red part of the L dwarf spectra (Fig. 10). To compute H_2O and HDO spectra we used line lists of Partrige & Schwenke [36]. Fortunately, the HDO bands should be shifted (in wavelength scale) relative to the H_2O bands. Indeed, HDO and H_2O are "light" molecules, – the ratio of their molecular wights $m_{H_2O}/m_{HDO} \sim 0.9$.

Let me note a few items:

— The shifts of HDO and H_2O bands are larger in the red part of the spectrum.

— Only in the case of high deuterium abundances (D/H $\sim 10-2$) we have a chance to detect "the heavy water" bands in the spectra of L dwarfs. Otherwise, the HDO bands disappear under H_2O bands.

The last result is of interest for the investigation of the evolution of the cool objects orbiting compact relativistic objects, which may have large deuterium (and lithium!, see [37]) abundances.

5 Conclusions

We arrive at the following conclusions.

- The processes of the formation of L dwarf spectra differ drastically from the case of M dwarfs. The overall spectral energy distributions of L dwarfs are governed by the pressure broadened wings of the resonance lines of potassium and sodium.
- The overall shape of optical spectra of the latest L dwarfs are governed by the K I and Na I resonance doublets; in the red we see molecular bands of CrH and FeH.
- The observed sequence of the L dwarf spectra is *the temperature sequence* (see [5]).
- L dwarfs spectra provide some unique possibilities of the verification of the existed methods of computations of profiles of the strongest resonance lines in cool plasma.
- Different phenomena: broadening by neutral species, NLTE, depletions into the dust particles, chromospheric-like features may affect the profiles of the strong lines.
- To model L dwarfs spectra we should understand the "dusty" properties of their atmospheres.
- The basic algorithm of the "lithium test" may be used for the assessment of the coolest L dwarfs and even T dwarfs.
- Bands of the "heavy water" are rather weak in the atmospheres of L dwarfs for the "cosmic" deuterium abundances, i.e. for D/H $\sim 10^{-5}$.

Acknowledgements

I'm grateful to R. Rebolo and M.R. Zapatero Ozorio for the fruitful collaboration and for providing the observational data in electronic form; to T.Tsuji for providing model atmospheres in digital form. I thank E. L. Martin for careful reading of the manuscript and for helpful comments.

Financial support of the project was provided by a Small Research Grant of the American Astronomical Society.

References

1. Martín, E. L., Basri, G., Delfosse, X., Forveille, T.: Astrom. Astrophys. **327**, L29 (1997)
2. Martín, E. L., Delfosse, X., Basri, G., Goldman, B., Forveille, T., Zapatero Osorio, M. R.: Astron. J. **118**, 2466 (1999)
3. *Kirkpatrick, J.D., Reid, I.N., Liebert, J. et al.* : Astrophys. J. **519**, 802 (1999).
4. Schultz, A.B., Allard, F., Champin, M., et al.: Astrophys. J. **492**, L.181 (1998).
5. Pavlenko, Y., Zapatero Osorio, M. R., Rebolo, R.: Astron. Astrophys. **355**,245 (2000).
6. Martín, E. L., Rebolo, R., Zapatero Osorio, M. R.: Astrophys. J. **469**, 706 (1996)
7. Kirkpatrick, J.D., Henry, J.T., McCarthy, D.W.Jr.: Astropys. J. Suppl. Ser. **77**, 417 (1991)
8. Allard, F., Hauschildt, P.H.: Astrophys.J. **445**, 433 (1995)
9. Pavlenko, Ya.V.: Astroph. Space Sci. **253**, 43 (1997).
10. Pavlenko, Y. V.: Odessa Astron. Publ. **10**, 76 (1997)
11. Pavlenko, Ya. V.: Astron. Reports **42**, 787 (1998)
12. Tinney, C. G.;Delfosse, X.; Forveille, T.;Allard, F.: Astron. Astrophys. **338**, 1066 (1998)
13. Kirkpatrick, J. D., Henry, T. J.,Liebert, J.: Astrophys. J. **406**, 701, 1993
14. Strauss, M. A., Fan, X., Gunn, J. E. et al.: Astrophys. J. Lett. **522**, L61 (1999)
15. Burgasser, A. J., Kirkpatrick, J. D., Brown, M. E., et al.: Astrophys.J. **522**, L65 (1999).
16. Tsuji, T.: ' Dust in very cool dwarfs'. In *"Low-Mass Stars and Brown Dwarfs in Stellar Clusters and Associations. La Palma"* eds. R.Rebolo, E.Martin, M.R.Zapatero Osorio., 43 (1998)
17. Auman, J.R., Woodrow, J.R.J.: Astrophys.J. **197**, 163 (1975)
18. Pavlenko, Ya,V.: *Effects of departure from LTE in atmospheres of M-giants* Tallinn, Valdus (1984)
19. Pavlenko, Y. V., Rebolo, R., Martin, E. L., Garcia Lopez, R. J.: Astron. Astroph. **303**, 807 (1995)
20. Tsuji, T.: Astron. Astrophys. **23**, 411 (1973)
21. Tsuji, T.: Astron. Astrophys. **305**, L1 (1996)
22. Gurvitch, L.V., Veitz, I.V., Medvedev, V.A. et al.: *Thermodynamic properties of the individual materials.* Moscow, Nauka 1979
23. Jones, H.R.A., Tsuji, T.: Astrophys.J. **480**, L39 (1997)
24. Pavlenko, Ya.V.: Astron. Reports **78**, accepted.
25. Kurucz, R.L.: Astrophys. J. Suppl. **40**,
26. Unsold, A.: *Physik der Sterntmospheren, 2nd ed.* Berlin, Springer (1949) 1.
27. Andretta, V., Comes, M.T., Severino, G.: Solar Phys. **39**, 19 (1991)
28. Nefedov, A.P., Sinelshchikov, A.V., Usachev,A.D.: Physica Scripta **59**, 432 (1999)
29. Burrows, A., Marley, M.S, Sharp, C.M.: Astrophys. J. **531**, 438 2000.
30. Pavlenko, Ya.V.: Astron.Reports **43**, 748 (1999).

31. Rebolo, R., Martín, E.L., Basri, G.W., Zapatero Osorio, M.R.: Astrophys. J. Lett. **469**, L53 (1996)
32. Martín, E. L.;Basri, G.; Zapatero Osorio, M. R.: Astron. J. **118**, 1005 (1999)
33. Rebolo R., Martín, E.L., Magazzu, A.: Astrophys. J. **389**, L83 (1992)
34. Bejar, V. J. S., Zapatero Osorio, M. R., Rebolo, R.: Astrophys. J. **521**, 671 (1999)
35. Saumon, D., Hubbard, W. B., Burrows, A., Guillot, T., Lunine, J. I., Chabrier, G.: Astrophys. J. **460**, 993 (1996)
36. Partrige, H., Schwenke, D.W.: J. Chem. Phys. **106**, 4618 (1997)
37. Martín, E. L., Rebolo, R., Casares, J., Charles, P.: Nature, **358**, 129 (1992)

Part II

Observations

Introduction: The Coolest Dwarfs – a Brief History

R.F. Jameson

Astronomy Group, University of Leicester, LE1 7RH, UK

Abstract. In 1994 we knew of no objects below the bottom of the main sequence. Now large numbers of brown dwarfs in clusters and the field have been discovered. This short paper presents a brief history of the devlopment of observations of such objects from the first discoveries of 1995 to the present state of play.

1 Discoveries

In August 1994, Chris Tinney organised a conference entitled "The Bottom of the Main Sequence and Beyond" in Garching [9]. This conference could be summed up by saying there was no 'beyond', i.e. no brown dwarfs or extra-solar planets had been discovered despite considerable efforts. Only one year later, 3 brown dwarfs were known - GL229B, Teide 1 and PPL 15.

Teide 1 [7] and PPL 15 were both discovered in the Pleiades, which for some years had been recognised as a promising hunting ground since it is a young cluster (only about 100 Myr old) and brown dwarfs are relatively bright when young. PPL 15 had previously been proposed as a brown dwarf by John Stauffer but it was the discovery of lithium in its spectrum that confirmed this. Brown dwarfs are of course failed low mass stars that do not burn hydrogen and below 0.06 $M_\odot$ (60 Jupiter masses, M_J) they don't even burn lithium. Lithium burns at much lower core temperatures than hydrogen. Since low mass stars and brown dwarfs are fully convective, destruction of lithium in the core rapidly removes it completely. Thus lithium is a test for brown dwarfs. Brown dwarfs of solar composition can have masses up to 75 M_J. If they are as young as the Pleiades their core temperature has not reached its ultimate value and so lithium has not yet disappeared. Both Teide 1 and PPL 15 show the lithium resonance line at 6707 Å.

In contrast to the cluster brown dwarfs, GL229B was discovered by looking for very cool companions to nearby faint M dwarfs by Nakajima et al. [6] using a coronagraph, a device for blocking out the glare of the primary star. The discovery of GL229B was hard work, Nakajima and colleagues observed a hundred stars before finding GL299B. It was worth it; GL229B was amazingly cool with an effective temperature of only about 1000 K and by any model had to be a brown dwarf. It is so cool that its atmosphere contains methane which absorbs strongly in the H and K bands making it blue in its $J - H$ and $H - K$ colours.

2 Surveys

1995 was also about the time three major new sky surveys started. The Sloan Digital Sky Survey (SDSS) aimed to survey the northern galactic latitudes at optical and near infrared wavelengths out to one micron the maximum wavelength of a CCD. The 2 Micron All Sky Survey (2MASS) started surveying the whole sky in the in the J, H and K infrared bands to limits of $J = 16$ and $K = 15$ and the DEep Near Infrared Survey (DENIS) aimed to map the southern sky, at I, J and K, to similar limits.

The stage was set, it was known very cool objects like brown dwarfs existed and the tools were in place to find them. Sure enough very cool objects started to be discovered in considerable numbers.

As new cool object were discovered it became necessary to extend the Harvard classification of stars beyond M. M dwarfs, up until then the coolest known stars, are classified by the strengths of their molecular bands. The molecules involved are principally TiO and the metal hydrides in the optical while H_2O vapour dominates the 1 to 2.5 μm infrared. The next spectral class was called L. Two L dwarf classification schemes have been suggested by Kirkpatrick et al.[2] and Martin(1999)[5]. The L dwarfs have effective temperatures ranging from 2200 K to about 1500 K. Indeed one star GD165B was dicovered much earlier [1] which could not be classified as an M dwarf. So GD165B, which is the companion star to a white dwarf, claims the honour of being the first L dwarf. Its age is about 5Gyr, estimated from the properties of its white dwarf companion, which puts it right on the boundary between mainsequence stars and brown dwarfs. L dwarfs may be brown dwarfs, if young, or the faintest mainsequence stars. Of course the ages of field stars are not easily measured, but the lithium test has shown that several L dwarfs are brown dwarfs.

In the L band the TiO and other molecular features disappear as the metal oxides condense out to form dust. The infrared is still dominated by H_2O vapour. One important question is where is the dust located? Is it below or in the photosphere. Also the dust may form clouds so the L dwarfs could have weather systems like the giant planets and might thus be variable objects. M dwarfs are well known for strong magnetic activity with variable chromospheres. Does this activity persist into the L dwarfs. In which case L dwarfs might be variable on two counts.

The L classification does not extend to cover objects as cool as GL299B with an effective temperature of about 1000 K. Further objects like GL299B were only discovered a year ago (2000). They are classified as T dwarfs. Like GL299B the T dwarfs have methane in their photospheres which absorbs at 1.65, 2.35 and 3.3 microns. This depresses their H and K band emission giving them progressively bluer colours in $J-H$ and $J-K$. Thus based on their infrared colours alone they were difficult to distinguish from hotter stars,which explains why they were slow to be discovered. They emit virtually no optical emission but are detectable in the Z band ($\lambda \sim 0.9\mu$m) and have very red $Z-J$ colours. The combination of H_2O and CH_4 absorption makes for maximum emission in the J band, despite their very cool temperatures, so their overall spectrum is even more non blackbody like than the L dwarfs.

3 The Future

The past few years has seen the emphasis of research on very low mass cool objects switch from star clusters to field objects. The advantage of field objects is that they are near and therefore brighter and more amenable to follow up observations. Thus they provide oportunities to extend our detailed knowledge of the L and T dwarfs.

However, progress has also continued with the clusters. Some 50 brown dwarfs are now known in the Pleiades and brown dwarfs have also been found in the very young theta Orionis and sigma Orionis clusters, Lucas and Roche [4] and Zapertero Osorio et al. [10]. These two papers are particularly exciting since they claim to have discovered brown dwarfs with masses below 10 M_J, which could almost be called free floating planets.

Clearly much exciting work remains to be done with all these new discoveries. In particular exploring the boundary between L and T dwarfs, investigating the variability of L dwarfs and searching for more brown dwarfs in order to determine luminosty and mass functions are all hot topics. These are some of the subjects of the papers presented in this section.

References

1. E.E. Becklin & B. Zuckerman, Nature, **336**, 656 (1988)
2. J.D. Kirkpatrick, I.N. Reid, J. Liebert, R.M. Cutri, B. Nelson, C.A. Beichman, C.C. Dahn, D.G. Monet, J.E. Gizis & M.F. Skrutskie, 1999, ApJ, **519**, 802
3. S.K. Leggett et al., ApJ, **536**, L35 (2000)
4. P.W. Lucas & P.F. Roche, MNRAS, **314**, 858 (2000)
5. E.L. Martin, AJ **118**, 2466 (1999)
6. T. Nakajima, B.R. Oppenheimer, S.R. Kulkarni, D.A. Golimowsk, K. Mathews & S.T. Durrance, Nature, **378**, 463 (1995)
7. R. Rebolo, M.R. Zapatero-Osorio & E.L. Martin, Nature, **277**, 129 (1995)
8. J.R. Stauffer, D. Hamilton & R. Probst, AJ, **108**, 155 (1994)
9. C.G. Tinney (ed.), The Bottom of the Mainsequence and Beyond (ESO Workshop), Springer (1995)
10. M.R. Zapertero-Osorio, V.J.S. Bejar, E.L. Martin, R. Rebolo, D. Barrado y Navascues, C.A.L. Bailer-Jones & R. Mundt, Science, **290**, 103. (2000)

Imaging and Spectroscopy of Hot (Young) "Ultracool" Companions

G. Schneider[1], P.J. Lowrance[2], E.E. Becklin[2], J.D. Kirkpatrick[3], P. Plait[4], S.R. Heap[5], E. Malumuth[6], R.J. Terille[7], C. Dumas[7], A.B. Schultz[8], B.A. Smith[9], A.J. Weinberger[2], and D.C. Hines[1]

[1] Steward Observatory, University of Arizona, Tucson, AZ 85750, USA
[2] Dept. of Physics and Astronomy, UCLA, Los Angeles, CA 90095, USA
[3] IPAC, California Institute of Technology, Pasadena, CA 91125, USA
[4] Advanced Computer Concepts, Inc., GSFC, Greenbelt, MD 20771, USA
[5] GSFC, Greenbelt, MD 20771, USA
[6] Raytheon ITSS, GSFC, Greenbelt, MD 20771, USA
[7] JPL, California Institute of Technology, Pasadena, CA 91101, USA
[8] Astronomy Programs, Computer Sciences Corp., Lanham-Seabrook, MD 20706, USA
[9] Institute of Astronomy, University of Hawaii, Honolulu, HI 96822, USA

Abstract. We report the initial results from a program of HST/STIS spectroscopy of "hot" low-mass (brown dwarf) companions to nearby young stars identified in our HST/NICMOS coronagraphic imaging survey. We present spectra of the likely 20 and 40 Jupiter mass companions to CD -33° 7795 and HR 7329, respectively, the binary companions to GL 577 (both, possibly, brown dwarfs), and the low-mass star GL 503.2B. We discuss the state of our recent follow-up observations of the putative companion to TWA 6. We add these companions to the very few already known, allowing the first steps to be taken toward answering many fundamental questions regarding the formation and evolution of multiple systems with components near and below the stellar/substellar boundary.

1 Introduction

The discovery and subsequent characterization of sub-stellar objects in stellar systems continues to be one of the key goals of contemporary observational astronomy. From the low-mass end of the mass-luminosity function (which is still very poorly constrained for $M_* < 0.2 M_\odot$) through the regime of brown dwarfs and into the domain of giant planets, the near and sub-stellar bestiary remains fertile ground for exploration. Knowledge of stellar and sub-stellar masses and luminosities at and below the ~ 0.08 $M_\odot$ H-burning limit is of fundamental importance in many inter-related areas such as the determination of the stellar mass function at the low-mass "end" of the main sequence, the theory of stellar evolution, the resolution of age/evolution issues and the origins of stellar and planetary systems. Yet, for *companions*, the transition region between very low-mass stars and giant planets is poorly understood and ill-observed (due in large part to their intrinsic faintness and close proximities to their primaries).

With the discovery and imaging of the brown dwarf companion to GL 229 [24], the very existence of such objects moved them from the realm of conjecture to observation. And, since the first detection of an extra-solar planet in the Jovian-mass regime, the companion to 51 Peg [21], continuing discoveries of 1–10 Jupiter mass companions to solar-like stars by indirect methods have revealed an unanticipated diversity in mass ranges, dynamical properties and primary-star characteristics [19].

These radial velocity surveys suggest that $\sim 5\%$ of main-sequence stars possess 0.5-8 Jupiter-mass companions at distances typically $< 3\mathrm{AU}$ from their primaries, but <1% of these stars have brown dwarf companions. (e.g., [20]). Speculations as to the cause of this "brown dwarf desert" (e.g., [11]) may be observationally biased as these surveys are currently insensitive to higher mass objects at distances from tens to hundreds of AUs because temporal baselines of decades to centuries are required to detect them.

Recently, however, many isolated "field" brown dwarfs have been discovered from large ground-based surveys such as the 2-Micron All-Sky and Sloan Digital Sky surveys and subsequently confirmed spectroscopically ([13]; [9]). And, brown dwarfs (or "free floating planets") have been turning up in proliferation in very young clusters associated with star-forming regions (e.g., [18]; [23]). Indeed, population statistics in both the field and young clusters suggest that single brown dwarfs, once considered somewhat exotic objects, may be as common as stars (e.g., [33]; [10]). Yet, the *companion* mass and number fractions of substellar objects, which occupy a niche in the mass function between stars and planets, remain essentially unknown. How common are they? Only a few brown dwarf companions to stars have thus far been found and confirmed ([26]; [5]; [16]). The disparity may simply be a contemporary observational selection effect or the formation processes for field vs. companion brown dwarfs may be intrinsically very different. Does the star formation process end abruptly for masses below the H-burning limit in multiple systems? Do companion brown dwarfs form in processes more like planets than stars (i.e., fragmentation vs. core collapse; e.g., [4])? And, how might the presence of a brown dwarf companion effect the evolution of a newly-forming solar system?

The unequivocal identification and characterization of companion substellar objects in young stellar systems not only helps to enumerate the population statistics, but will ultimately lead to follow-up dynamical astrometric studies to ascertain their masses directly and unambiguously. Though many such objects will have to be discovered and studied before questions such as those posited above are fully answered, with those now in hand, we have a logical starting point to begin investigating the intriguing possibilities.

2 Observations

Long before the adoption of NASA's *Origins* theme, which contains in its foundation the establishment of scientific activities to investigate the birth and early evolution of stars and planets [8], the NICMOS Instrument Definition Team embraced these goals in formulating our Environments of Nearby Stars (EONS) program. A cornerstone of our EONS investigations was a systematic search for previously unknown low-mass candidate companions to young and nearby stars. Carried out during HST Cycle 7, this survey was designed to find companion candidates spanning two decades in the mass-spectrum down into the regime of Jovian mass planets (for objects as young as $\sim$ 10 Myr). Using the Near-Infrared Camera and Multi-Object Spectrometer (NICMOS) on the Hubble Space Telescope (HST), we conducted a coronagraphic imaging survey of young and nearby stars searching for, and finding, candidate brown dwarf and giant planet companions. We have now confirmed the physical association of a subset of those objects through the establishment of common proper motions of the candidates with their putative primaries. Further, we have determined their spectral types and photospheric temperatures using near-IR/optical spectra obtained with the Space Telescope Imaging Spectrograph (STIS).

2.1 NICMOS Imaging – Detection

Approximately $\sim$ 18% of the NICMOS IDT's observing time on HST was dedicated to NICMOS/EONS survey, as previously discussed in detail in the literature ([27], [14], [29]). In summary, we coronagraphically imaged the circumstellar regions of eighty-three young and nearby stars, primarily in H-band. In particular we observed:

1) Thirty-eight very young main-sequence stars with mean distances of $\sim$30pc. Young sub-stellar companions will still be in a higher luminosity phase and thus more readily detectable. The median age for our candidates with well-established ages is $\sim$ 90Myrs with 8 candidates as young, or younger than, 10Myrs. This sample includes several members of the TW Hydrae association (d $\sim$ 55pc), the nearest site of recent star-formation to the Earth [12].

2) Twenty-seven M-dwarfs which are a) very nearby (d < 6pc), with spectral types later than $\sim$M3.5; b) young (age < 10 Myr) and (d < 25 pc); and c) spectrally the latest known (i.e., "ultra-cool" dwarfs later than $\sim$M8.5).

3) Eighteen primarily main-sequence stars with large IRAS excesses and other indicators of dust, to image circumstellar material at 1.1 and/or 1.6μm.

Each target was observed in a single orbit (providing a total of $\sim$ 25 minutes of integration time), but at two different spacecraft orientation angles to permit the discrimination of point spread function artifacts from intrinsic circumstellar sources. By subtracting two coronagraphically occulted images of a star, rolled about the target axis, the NICMOS/HST system can achieve a per-pixel reduction in background intensity approaching 10^{-6} of the peak stellar intensity at a distance of 1". In H-band this allowed us to achieve statistically significant detections of companions as faint as H $\sim$ 24, 10–15 magnitudes fainter than their primaries at separations of 0.5"–3.0" [28]. For the youngest objects in our sample these levels of performance enable the imaging of "hot" Jupiters at distances of tens of AUs from their primaries. Typically, the stability of the instrumental system allows us to determine the positions of coronagraphically imaged putative companions with a precision of about 10mas with respect to their suspected primaries in unocculted target acquisition images, establishing an astrometric baseline for follow-up observations.

2.2 Ground-Based (AO) Imaging – Confirmation

The unequivocal determination of companionship for a candidate rests in the confirmation of physical association with its putative primary through the establishment of common proper motions. A subset of our targets, those with less severe contrast ratios and/or larger separations, are observable with ground-based adaptive optics systems. With our first-epoch NICMOS observations, and later ground-based AO imaging which we are obtaining on the Palomar, Keck, and CFHT AO systems (for those observationally possible), we are able to discriminate against background objects for many of these targets on relatively short time-scales. For several high proper motion objects such astrometric confirmation of companionship has already been secured. For four astrometrically confirmed objects, of six thus far also studied spectroscopically, we have found CD -33° 7795B, HR 7329B, and GL 577B/C to be brown dwarfs (the latter a brown-dwarf binary, the first such hierarchical companion found), and GL 503.2B to be a very-low-mass star near the stellar/brown dwarf boundary. The astrometric results for TWA 6"B", an extrasolar giant planet (EJP) candidate, are indeterminate at this time, as a longer temporal baseline is required to confirm or reject its companionship.

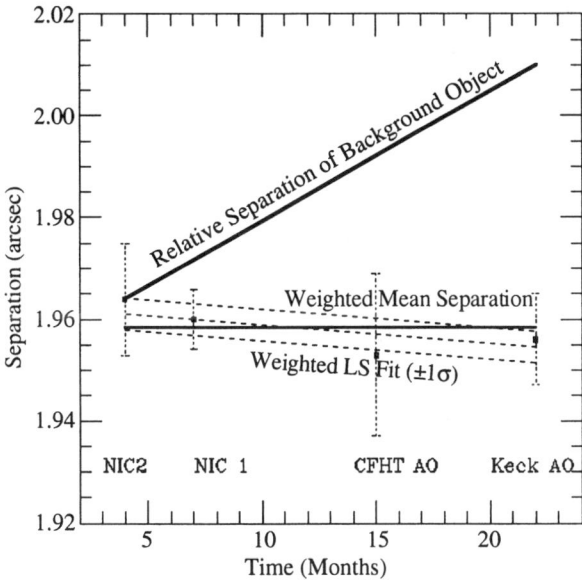

Fig. 1. Companionship is confirmed (or rejected) by differential proper motions. As an example we show angular separation measures for CD -33° 7795A/B over an 18 month baseline which, statistically, remained constant (within the measurement error). We compare this with the anticipated change in separation from the proper motion of CD -33° 7795A if the companion was actually background star, and conclude they are indeed a common proper motion pair.

2.3 STIS Spectroscopy – Characterization

During HST Cycle 8 we obtained STIS spectra for a subset of our young (less than ~ 0.3 Gyr) NICMOS candidate companions to find their spectral types and effective temperatures. The instrumental characteristics of STIS+HST are uniquely suited to the task of characterizing the substellar nature of these targets. The ability of STIS to obtain spectra in the regions of key atomic lines and molecular bands to differentiate between very low mass stars, brown dwarfs, and extra-solar giant planets, for very faint companions in close angular proximity to their bright primaries, is unmatched by any current ground-based capabilities. Representative spectra of two young objects CD -33° 7795B and HR 7329B,

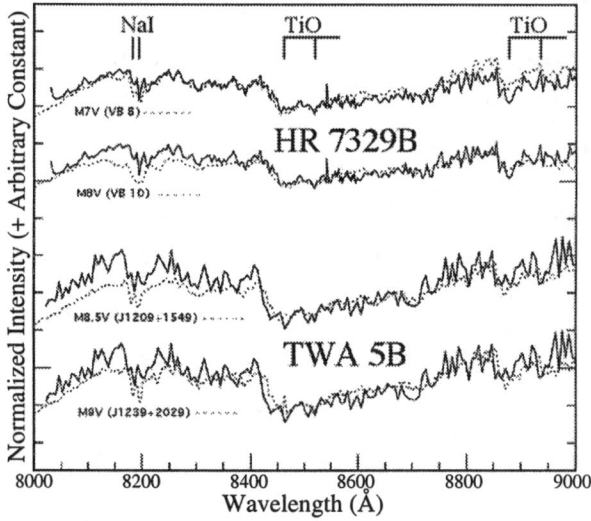

Fig. 2. Young and therefore hot, brown dwarfs have spectra resembling M dwarf stars of the same photospheric temperatures, rather than exhibiting spectral characteristics of older L and T dwarfs. Here we compare our STIS spectra of TWA 5B (CD - 33° 7795B)and HR 7329B, both presumed to be one to a few x 10^7 yrs of age, with several M stars. We ascertain their spectral types on the basis of the continuum, the strength of the NaII doublet and the depths of the TiO band-heads.

in the 8000–9000Å range, are compared to several late M-dwarf standards of similar effective temperatures in Fig. 2. In the case of CD -33° 7795B we find a very good fit for M9V and an effective temperature of 2600K. For HR 7329B we deduce a spectral class of M7.5V and effective temperature of 2700K. STIS spectra in $\sim$ 7500–9500Å range allows us to compare the strengths and morphologies of diagnostic spectral tracers in late M, L, and T dwarfs (and giant planets). Young, and therefore hot, brown dwarfs have spectra resembling M dwarf stars of the same photospheric temperatures. In this spectral region, M dwarfs are dominated by bands of TiO and, for the latest M dwarfs, VO. For cooler L-types, the oxide molecules disappear onto grains leaving cleaner spectra showing only bands of the hydride molecules CrH 8611Å and FeH 8692Å along with sharp atomic lines of the alkali metals Na 8183 & 8195Å, Rb 7800 & 7948 Å, and Cs 8521 & 8943Å. In the spectra of the T class, which abut the L dwarfs in temperature, the hydrides have disappeared onto grains and some of the alkali metals have formed into chloride compounds, leaving Rb and Cs lines (as in GL 229B) and, for cooler spectra, CH_4 8890Å. H_2O vapor bands ($\sim$ 9300–9600Å), which gain strength throughout the sequence from late-M through the methane types, are significantly compromised in ground-based spectra due to atmospheric absorption (e.g., see figure 4 of [30]).

STIS target acquisition and dispersed spectral images also allow us to constrain further the relative proper motions of the presumed systemic components. For example, a spectrum was secured for HD 102982B [14] in the 0.2" wide slit, which would have failed for most possible geometries if it was a background object. Conversely, in the case of HD 177996 companionship for the candidate was rejected as the object did not appear in the slit (because of its differential proper motion as a background star) after the "primary" was successfully acquired.

3 Hot Brown Dwarf Companions

While the spectra of objects like CD -33° 7795B and HR 7329B clearly resemble late M dwarfs, their youth belies their true nature - "hot" brown dwarfs. A fundamental, and reasonable, assumption is that the low-mass companions are coeval with their primaries. With effective temperatures determined from our spectra, and bolometric magnitudes derived from our H-band imaging, we can then place the companions on H-R diagrams demanding coevality with their primaries, and estimate their masses through evolutionary cooling models.

3.1 CD -33° 7795 (TWA 5)

CD -33° 7795A, an M1.5 dwarf (H=7.2) is a member of the member of TW Hydrae Association [12]. With an estimated age of 10 Myrs ([12]; [32]) the primary exhibits X-ray and H-α emission, and strong Li absorption. CD -33° 7795 was a suggested spectroscopic binary, so chromospheric activity may be enhanced [3], but its space motions and Li absorption are consistent with other TW Hydrae association members. Keck AO imaging resolved the primary as a close ($\sim$ 0.06") binary with nearly-equal magnitude components (see Fig. 3). However, the close binary orbit must be in a nearly edge-on as it was unresolved at the time of the NICMOS observations.

CD -33° 7795B was the first brown dwarf companion candidate identified in the NICMOS survey [15], and co-discovered by [34] in speckle observations taken at the NASA IRTF. An H=12.14 magnitude companion to TWA 5A (ΔH=4.92) was readily seen in the NICMOS F160W (1.6μm) coronagraphic difference image with a separation of 1.96" ($\sim$ 110 AU projected distance) at a PA of 358.9°. Near-IR colours were measured from KeckII/LRIS acquisition imaging, possible given the modest primary/secondary contrast ratio and sufficiently large separation.

62 G. Schneider et al.

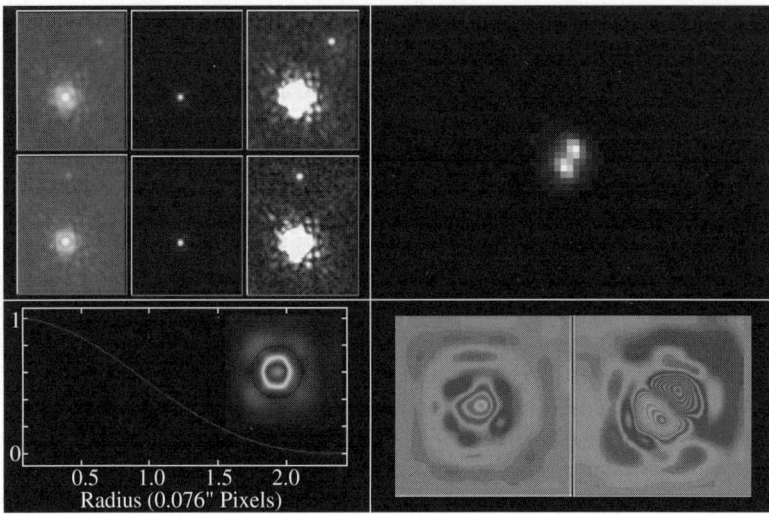

Fig. 3. Direct imaging of CD -33° 7795A. Top Left - NICMOS 1.65µm target acquisition (TA) images at two spacecraft orientations (three stretches to show PSFs of A and B components). Top Right - Keck 1.6µm AO image, 22 months later, revealing the primary as two equal magnitude components separated by 60mas. Bottom Left - Combined NICMOS TA image and radial profile of A component shows no evidence of duplicity. Bottom right - PSF core of A component fell at different sub-pixel locations in the two rolled NICMOS TA images. Difference of the two TA images (left), stretched to show residuals from imperfect subtraction, bears no resemblance to the difference image of two equal magnitude point sources separated by 60mas (right).

Differential astrometric measures from HST/NICMOS, CFHT AO and Keck AO images over a baseline of 18 months confirmed companionship for the secondary through common proper motions (see Fig. 1). From the NICMOS and Keck photometry, and a bolometric correction of BC(H) = 2.8, we ascertained its luminosity as 0.0021 $L_\odot$ and its effective temperature as $\sim$ 2600K. From these measures we determined a photometric spectral classification of M8.5V ($\pm$ 0.5 spectral class). These results were confirmed by the STIS spectra (see Fig. 2), though a better fit to the template spectra is obtained at M9V (still within the estimated photometric uncertainty). Recently [25] obtained lower resolution spectra using FORS2 on the VLT which are in agreement with these results.

3.2 HR 7329

HR 7329A, a member of the recently discovered Tucanae association [35] is of comparable distance (d = 47.61 $\pm$ 1.6pc as determined by the Hipparcos mission) and age as CD -33° 7795. Though somewhat uncertain, it is likely to be 10–30 Myr, based upon its location on a HR diagram, very large Vsini (330 km/sec; [1]), and Tucanae association membership. A brown dwarf companion candidate was identified with NICMOS, in close proximity to the H=5.05 A0V primary

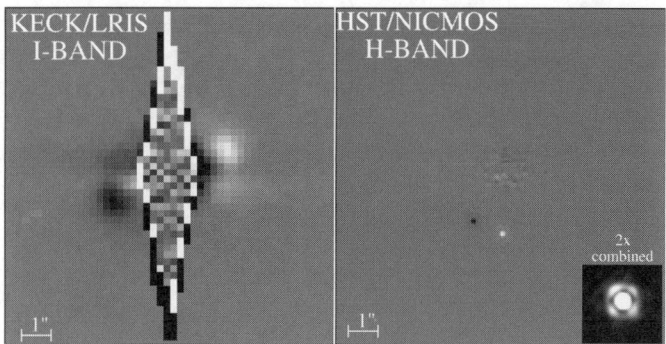

Fig. 4. Left - Keck/LRIS I-band image of CD -33° 7795, self-subtracted after 180° image rotation shows (positive and negative) images of B component. Right - NICMOS H-band coronagraphic images CD -33° 7795B, spacecraft target point rolled around component A and subtracted. Inset - positive and negative image conjugates from subtraction combined.

at an angular distance of 4.17" [16]. Again, assuming physical association for the unresolved object (196 AU projected distance), the H=11.90 companion candidate at PA of 166.8° was estimated to have a luminosity of 0.0026 L$_\odot$ (Habs = 8.54 with BC(H)=2.67).

3.3 GL 577

While outside of the NICMOS coronagraphic aperture, a faint "extended" object was seen in two short NICMOS target acquisition images of the young star GL 577 [22] (H=6.88, G5V) at a distance of 5.35" (~240 AU projected separation). Using the ~50x brighter image of GL 577A as a reference PSF, the "extended" object was nulled by two point sources differing by <15% in H-band brightness separated by ~0.1". The binary nature of this suspected companion was recently confirmed with follow-up imagery obtained with the PHARO adaptive optics camera on the Palomar 200" telescope [17]. Companionship (thus coevality) for the GL 577B/C pair was established by their common proper motions with GL 577A over the ~ 2 year baseline since the NICMOS observation.

A blended spectrum of GL 577B/C (see Fig. 7), recently obtained with STIS, resembles that of an M5.5V star. Differences in detail arising from the superposition of the individual spectra are noted and careful deblending of the spectra will be required to ascertain properly the spectral types and effective temperatures of the components. However, both components are roughly equal in luminosity (derived from the spectral type, H magnitude and distance) and they each have effective spectral temperatures of ~ 2900K.

3.4 Mass Estimates

Brown dwarfs cool as their luminosities decline with age, approximately as $L \sim M^{-2.24}T^{-1.3}$. Hence, mass estimates depend upon the efficacy of the age

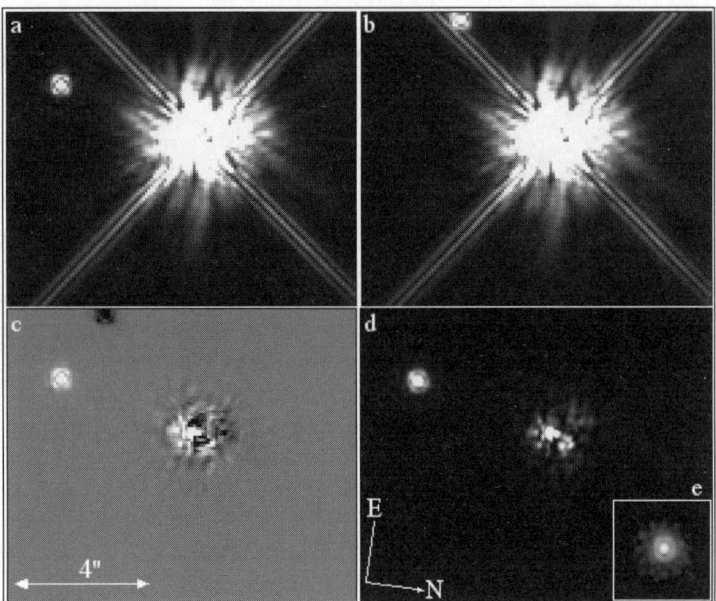

Fig. 5. Top (a,b) - NICMOS coronagraphic images of HR 7329A & B. In panel (b) the companion PSF is partially off of the detector. (c) - Difference image (rolled 29.9° about primary component). (d) - Combined difference image conjugates. Inset (e) - Detailed recovery of roll combined PSF.

3.4 Mass Estimates

Brown dwarfs cool as their luminosities decline with age, approximately as $L \sim M^{-2.24}T^{-1.3}$. Hence, mass estimates depend upon the efficacy of the age determinations, and of the details of cooling models themselves. For older stars determining the ages of the more massive primaries becomes more difficult and uncertain. Obtaining independent age estimates for the newly determined companions via, for example, Lithium abundance, will be a very important check and constraint for these mass estimates. In all cases considered here, we currently rely on the ages determinations of the primaries.

Assuming coevality with their respective primaries, we place the companions on evolutionary tracks (e.g., Fig. 8) from [2], [6], and [7] to estimate their masses. For CD -33° 7795B we find a mass of $\sim$ 20 Jupiters, and for HR 7329B $\sim$ 40 Jupiters. In the case of GL 577B and C, we assume a B/C bolometric luminosity ratio of at most 1.15:1, implicating both as transitional objects, with one, and likely both brown dwarfs.

3.5 Low Mass, but Not Low Enough

With an estimated age of 300 Myrs for the H=5.45 solar-like (G2V) UMa group primary GL 503.2A (n(Li)=2.56, EW=77Å, vsini=25km/s; [31]), the $\sim$ M4.5V

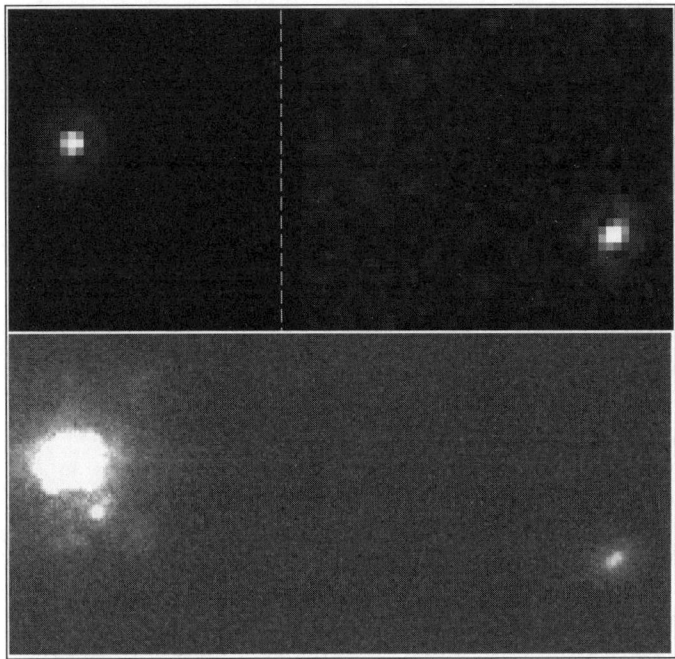

Fig. 6. Top - NICMOS TA image of GL 577A & B/C (flux renormalized x50 for B/C companions on right side of frame). Bottom - BOTTOM: Palomar 200" AO image (bright spot below-right of primary is a filter ghost).

spectrum of its companion is indicative of a low mass star rather than a hot brown dwarf (Fig 7, top). Thus, we reject the possibility of the sub-stellar nature of this 1.56" distant companion (projected separation of 40 AU at d = 25.7 pc.) With an estimated effective temperature of 3150 ± 100K, the companion mass is found to be in the range of ∼ 0.1 to 0.15 solar masses.

4 Extrasolar Giant Planets?

4.1 Imaging

The feasibility of imaging young Jovian-mass planets by NICMOS coronagraphy was demonstrated with the acquisition of a S/N=50 1.6μm image of a candidate extra-solar giant planet companion to ROSAT 116 (H=6.9, K7V), also a member of the TW Hydrae association (TWA 6). The H=20.1 putative companion seen in Fig. 9 is 2.5" (∼ 140 AU projected distance) from and 13.2 magnitudes fainter than TWA6. In a 0.9μm NICMOS camera 1 observation the object was not seen to a limiting magnitude of > 22, indicative of a very red source (if not a highly reddened background object). *If* the object is physically associated and coeval with ROSAT 116, its absolute H magnitude suggests an effective surface

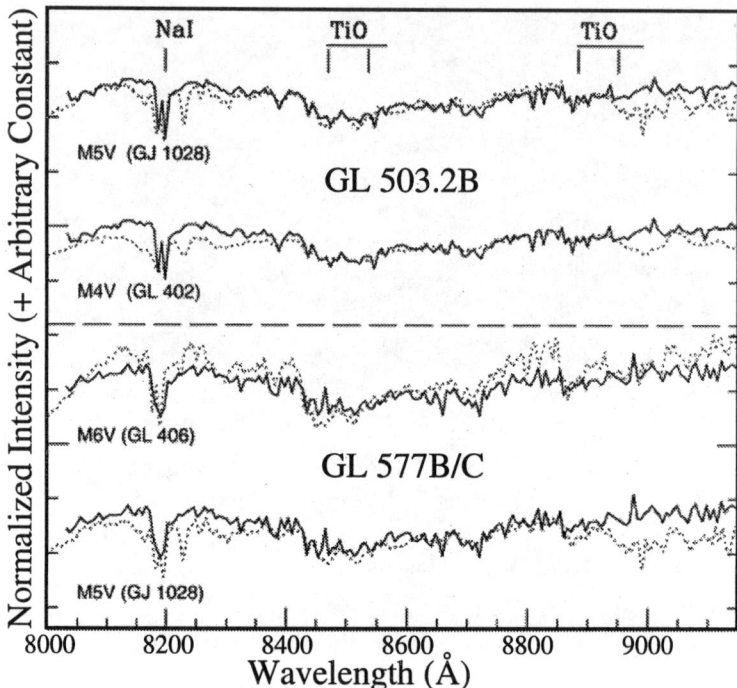

Fig. 7. Spectrum of GL 503.2B (top) and blended spectra of the B/C companions to GL 577 (bottom) compared to those of late M dwarfs (dotted lines). Effective photospheric temperatures of ~ 3150K for GL 503.2 and ~ 2900K for each component of ~ 2900K is found.

temperature of ~ 800K (for a surface gravity of 7.5×10^4 cm^2 s^{-1}) and a mass of ~ 1 Jupiter based upon evolutionary models updated from [6].

We are attempting to use the Keck AO system to establish (or reject) the physical association of this candidate with TWA 6. Currently, with less than a two year baseline, the measurement errors in the relative proper motions are of the same order of magnitude as the proper motion of TWA 6 itself. Further astrometric calibration (and a longer temporal baseline) is required to improve the relative astrometry which, at the moment, is inconclusive.

4.2 Spectroscopy

Because of the extreme contrast ratios and close angular proximities of EJP companions, spectroscopy is much more of a challenge than imaging. We recently obtained a low resolution STIS spectrum of TWA 6'B' in six HST orbits. Not unexpectedly, its faint spectrum is highly contaminated by the dispersed light of the much brighter primary, which is in close proximity to the slit. A "spray" of scattered light in between the companion spectrum from those arising from the diffraction spikes of the primary is due to instrumental scattering

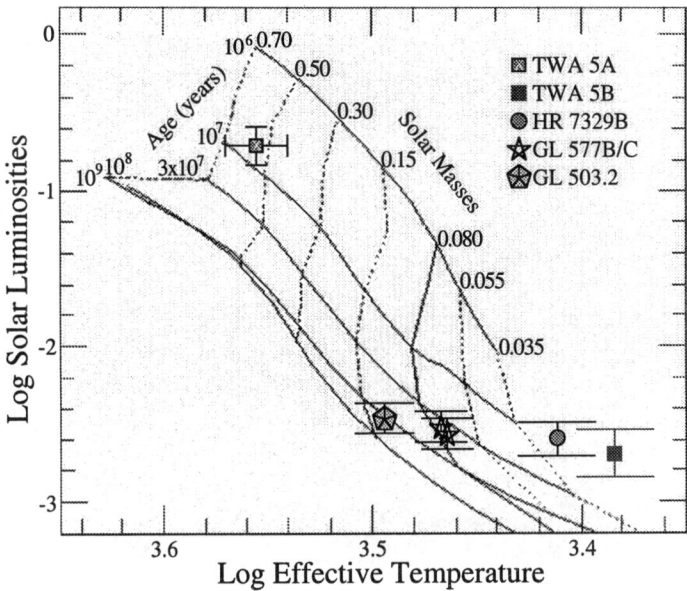

Fig. 8. Placement of companions on an H-R diagram and in the context of evolutionary isochrones of low-mass stars and substellar objects [2] based upon luminosities derived from NICMOS imaging and effective temperatures inferred from STIS spectra.

and diffraction effects. To date attempts at modeling and removing the background contamination have been only partially successful. Additional work is required to understand better the instrumental scattering and the limits imposed by the STIS optical system before a photometrically calibrated spectrum of the candidate object may be properly extracted. Scattered light calibration observations are scheduled for late 2000. Unfortunately, at this juncture, any interpretation would be premature.

5 Summary

The NICMOS coronagraphic survey, follow-up ground-based AO imaging and STIS spectroscopy, have given rise to the discovery, confirmation, and preliminary characterization of several very low-mass and sub-stellar companions. Here we have reported on the first results from this survey. Additional candidates identified by NICMOS are being followed up in a similar manner. By assuming coevality with their primaries, we can estimate the masses of the companions through evolutionary cooling models (e.g., see Fig. 10, adapted from [6]). Age determinations of the primaries, however, are somewhat uncertain (particularly for the older stars in our sample). Hence, independent determinations of the ages of the companions (e.g., via Lithium abundance and H-α equivalent widths) is

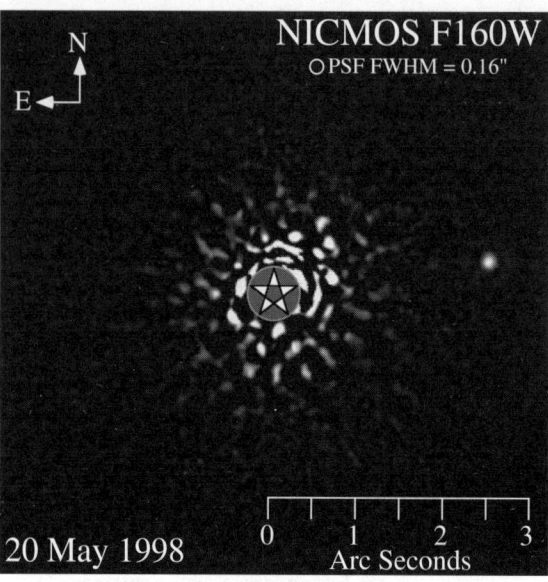

Fig. 9. An EJP companion *candidate* to ROSAT 116 (TWA 6) is readily seen in this NICMOS F160W coronagraphic image. Though the nature of the object is still to be determined, the viability of direct detection of young EJPs via NICMOS differential coronagraphy is demonstrated by this image.

essential to constrain better the nature and properties of the newly-discovered companions.

Acknowledgements

This work is based, in part, on observations with the NASA/ESA Hubble Space Telescope obtained at the Space Telescope Science Institute, which is operated by AURA, Inc. under NASA contract NAS5-26555 and supported by grants NAG5-3042 and GO-98.8176A to the NICMOS Instrument Definition and Environments of Nearby Stars teams. We acknowledge additional support by NASA grant NAG5-4688 to UCLA and the use of the Palomar and W.M. Keck observatories; the latter operated as a scientific partnership between the California Institute of Technology, the University of California and NASA.

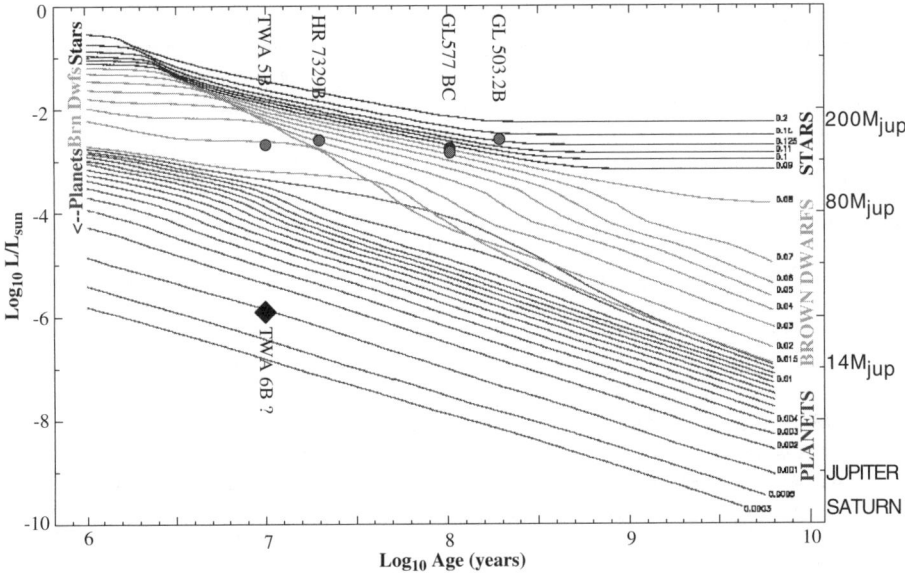

Fig. 10. With ages assumed from the characteristics of their primaries we estimate masses for the well-established companions observed spectroscopically with STIS (circles), based upon evolutionary cooling models for very low mass stars and substellar objects from [6]. The classification of "transitional" objects (such as GL 577B/C and possibly GL 503.2B), and the characterization of all objects during their cooling phases depends critically upon their presumed ages. Intrinsically fainter objects, such as the possible companion to TWA 6 (diamond) may be observed into the planetary-mass domain for objects of sufficient youth.

References

1. H.A. Abt, N.I. Morrell, ApJ, **99**, S135 (1995)
2. I. Baraffe, G. Chabrier, F. Allard, P.H. Hauschildt: A&A. **337**, 403 (1998)
3. D. Barrado Y Navascues, J.R. Stauffer: A&A. **310**, 879 (1996)
4. A.P. Boss: 'Modes of Gaseous Planet Formation'. In: *Planetary Systems in the Universe*, IAU Sumposium 202, August 2000.
5. A.J. Burgasser et al.: ApJ **531**, L57 (2000)
6. A. Burrows et al.: 'The Spectral Character of Giant Planets and Brown Dwarfs'. In: *ASP Conf. Ser. 154, The Tenth Cambridge Workshop on Cool Stars, Stellar Systems and the Sun* ed. by R. A. Donahue and J. A. Bookbinder, p.27 (1998)
7. F. D'Antona, I. Mazzitelli: Mem. Soc. Astron. Italiana, **68**, 807 (1997)
8. A. Dressler: *HST and Beyond: Exploration and the Search for Origins*, (AURA, Washington, DC 1996).
9. X. Fan et al.: AJ **119**, 928 (2000)
10. J.E. Gizis. D.G. Monet, I.N. Reid, J.D. Kirkpatrick, J. Liebert, R.J. Williams: AJ **120**, 1085 (2000)
11. J.L. Hallwachs, F. Arenou, M. Mayor, S. Udry, D Queloz: A&A **355**, 581, 2000
12. J.H. Kastner, B. Zuckerman, D.A. Weintraub, T. Forveille: Science **227**, 67 (1997)

13. J.D. Kirkpatrick et al.: AJ **120**, 447 (2000)
14. P.J. Lowrance et al.:'A Coronagraphic Search for Substellar Companions to Young Stars'. In: *NICMOS and the VLT: A New Era of High Resolution Near Infrared Imaging and Spectroscopy, Pula, Sardinia, Italy, May 26-27, 1998* ed. by W. Freudling, R. Hook (ESO, Garching, 1998), pp. 96-101.
15. P.J. Lowrance et al.: ApJ **512**, 69 (1999)
16. P.J. Lowrance et al.: ApJ **541**, 390 (2000)
17. P.J. Lowrance et al.: ApJ submitted (2000)
18. K.L. Luhman et al.: ApJ **540**, 1016 (2000)
19. G.W. Marcy, W.D. Cochran, D. Mayor:'Extrasolar Planets Around Main-Sequwnce Stars'. In: *Protostars and Planets IV.* ed. by V. Mannings, A.P. Boss, S.S. Russell (University of Arizona Press, Tucson, 2000), pp. 1285-1311
20. G.W. Marcy, P.R. Butler: PASP, **112**, 137 (2000)
21. M. Mayor, D. Queloz: Nature **378**, 355 (1995)
22. S. Messina, E.F. Guinan, A.F. Lanza: Ap&SS (260), 493 (1999)
23. R. Mundt, C.A.L. Bailer-Jones,M.R. Zapatero Osorio, V.J.S. Bejar, R. Rebolo, D. Barrado Y Navascues, E.L. Martin:'Discovery of Very Young Free-floating Giant Planets in the Sigma Orionis Cluster'. In: *Astronomische Gesellschaft Abstract Series, Vol.17. Abstracts of Contributed Talks and Posters at the Annual Scientific Meeting of the Astronomische Gesellschaft at Bremen, September 18-23, 2000* pp 11.
24. T. Nakajima, B.R. Oppenheimer, S.R. Kulkarni, D.A. Golimowski, K. Matthews, S.T. Durrance: Nature **378**, 463 (1995)
25. R. Neuhäuser, E.W. Guenter, M.G. Petr, W. Brandner, N. Huèlamo, J. Alves: A&A **360**, L39 (2000)
26. B.R. Oppenheimer, S.R. Kulkarni, K, Matthews, T. Nakajima: Science **270**, 1478 (1995)
27. G. Schneider,:'NICMOS Coronagraphic Surveys: Preliminary Results'. In: *NICMOS and the VLT: A New Era of High Resolution Near Infrared Imaging and Spectroscopy, Pula, Sardinia, Italy, May 26-27, 1998* ed. by W. Freudling, R. Hook (ESO, Garching, 1998), pp. 88-95
28. G. Schneider, R.I. Thompson, B.A.Smith, R.J. Terrile: 'Exploration of the Environments of Nearby Stars with the NICMOS Coronagraph: Instrumental Performance Considerations'. In: *Space Telescopes and Instrumentation V., Proc. of the SPIE*, Vol. *3356* ed. by P. Bely, J. Breckinridge (SPIE, Kona 1998), pp. 215
29. G. Schneider, E.E. Becklin, P.J. Lowrance, B.A. Smith: 'Substellar Companions to Nearby Stars from NICMOS Surveys'. In: *Disks, Planets and Planetesimals, ASP Conference Series* ed. by F. Garzon, C. Eiroa, D. de Winter, T.J. Mahoney (ASP 2000) in press.
30. A.B. Schultz et al: ApJ **492**, L181 (1998)
31. D.R. Soderblom, J.R. Stauffer, K.B. MacGregor, B.F. Jones: AJ **105**, 2299 (1993)
32. D.R. Soderblom et al: ApJ **498**, 385 (1998)
33. Z.I. Tsvetanov et al.: ApJ **531**, 61 (2000)
34. R.A. Webb, B. Zuckerman, I. Platais, J. Patience, R.J. White, M.J. Schwarz, C. McCarthy: ApJ **512**, L63 (1999)
35. B. Zuckerman, R.A. Webb: ApJ **535**, 959 (2000).

Activity and Kinematics of M and L Dwarfs

J.E. Gizis

Infrared Processing and Analysis Center, California Institute of Technology, Pasadena, CA 91125, USA

Abstract. I discuss observations of two traditional age indicators: chromospheric activity and kinematics, in late-M and L dwarfs near the hydrogen-burning limit. The frequency and strength of chromospheric activity disappears rapidly as a function of temperature over spectral types M8-L4. There is evidence that young late-M and L dwarfs have weaker activity than older ones (the opposite of the traditional stellar age-activity relation). The kinematics of L dwarfs confirm that lithium L dwarfs are younger than non-lithium dwarfs.

1 Introduction

The Two Micron All-Sky Survey (2MASS) has enabled large samples of cool dwarfs to be detected and studied. Over the spectral range M8 to L8, the traditional TiO and VO molecular bands disappear, grains become important, and the appearance of the spectrum changes drastically [13].

This spectral range, corresponding to temperatures between ~ 2300 and ~ 1400K, covers the crucial transition between hydrogen-burning stars and brown dwarfs. The situation is illustrated in Figs. 1 and 2 using theoretical models [6] and a plausible temperature scale [13]. (See this volume for different views of the L dwarf temperature scale). Both stars and brown dwarfs cool as they contract. Stars, however, eventually stabilize at a constant temperature (the upper tracks in Fig. 1), while brown dwarfs continue to cool (the lower tracks). Current theoretical models suggest that $0.075 M_\odot$ stars may exist as cool as ~ 1800K [6,1], corresponding to $\sim$L4. (In the hotter L dwarf temperature scales, even later L dwarfs may be stars.)

On the basis of Fig. 1, it is clear that field brown dwarfs and stars could be easily distinguished if we could measure age – but age is not directly measurable and must be deduced from indirect measures. Luminosities (in the form of absolute magnitudes) and temperatures (in the form of colours and spectral types), *can* easily be measured, but as seen in Fig. 2, different mass objects follow evolutionary tracks that are practically indistinguishable.

The need for an age indicator for M and L dwarfs motivates an investigation of two traditional stellar age indicators. Chromospheric activity is linked to age in convective stars through the dynamo: a star is born rapidly rotating, the rotation drives a dynamo which produces magnetic fields, the fields produce activity (chromosphere, corona, flares, and wind) the wind spins down the star, the dynamo is weakened, and the observable activity then decreases with age.

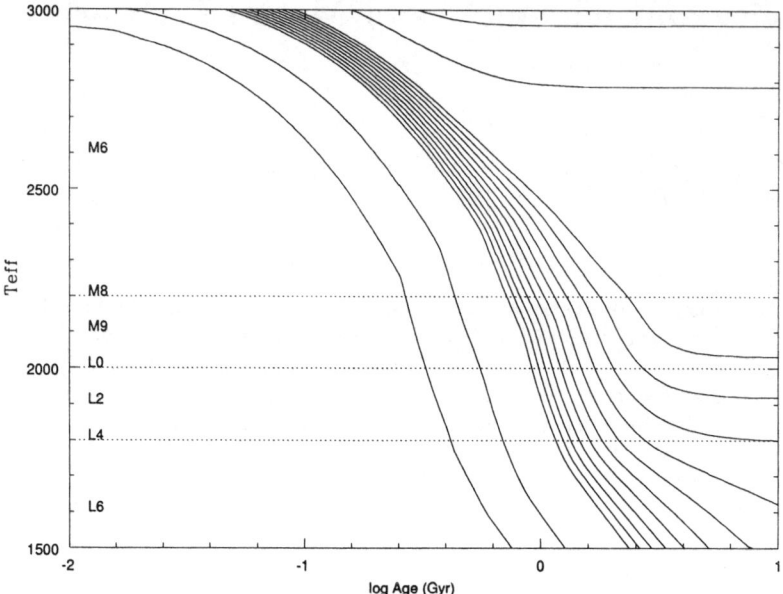

Fig. 1. Theoretical model tracks for stars and brown dwarfs [6]: Age information allows stars and brown dwarfs to be distinguished.

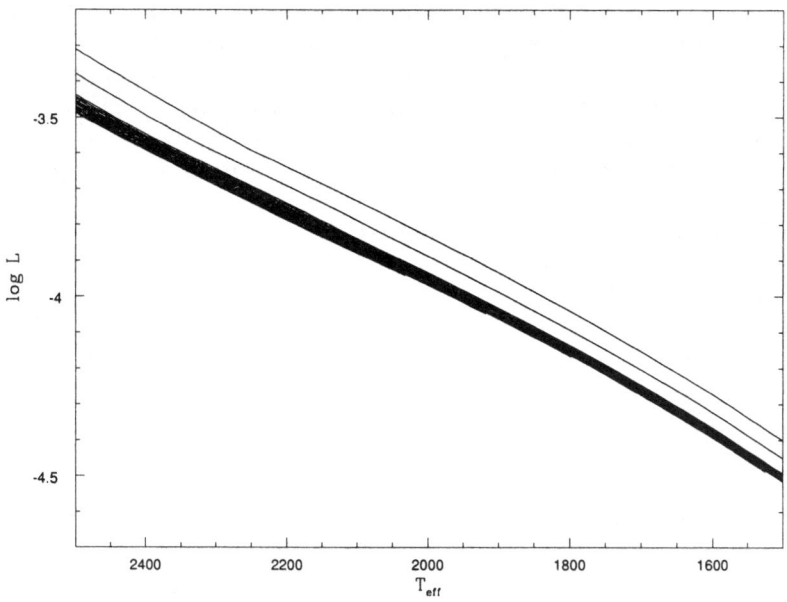

Fig. 2. Theoretical model tracks for stars and brown dwarfs [6]: Temperatures and luminosities do not provide mass or age information.

Kinematics are also linked to age: Stars are born from molecular clouds with low random space velocities, but with time encounters other stars and clouds in the Galactic disk 'heat' the stellar velocity distribution. Most of the discussion in this paper is drawn from Gizis et al. [10].

2 M and L Dwarf Observations

Two large samples form the basis of the discussion in this paper. Late-M and early-L dwarfs have been selected to $K_s < 12$ on the basis of their colours without any kinematic bias [10]. This sample is supplemented by the large sample of fainter L dwarfs also selected on the basis of colour [14]. Observationally, a sample of nearby bright M and L dwarfs offers considerable advantages. They are bright enough to relatively easily obtain far-red spectroscopy (> 6000Å), allowing the measurement of spectral types, Hα emission, lithium absorption, surface gravity, and a photometric parallax (Figs. 3, 4, and 5). This can be supplemented by astrometry, providing proper motions, trigonometric parallaxes, and hence tangential velocities. In the case of the late-M dwarfs, the sample is bright enough that the Palomar plates plus 2MASS allow measurements of the proper motions, which Gizis et al.[10] combined with the photometric distance estimate to obtain v_{tan}. The Kirkpatrick et al.[13,14] L dwarfs are faint enough that this is not possible, but United States Naval Observatory has measured proper motions and trigonometric parallaxes for a representative sample using CCD astrometry. Together, near-infrared photometry, spectroscopy and astrometry provide a rich set of diagnostics. The discovery of additional bright L dwarfs with $K_s < 12$ will aid many studies, such as 3.3 micron observations of methane (Geballe, this volume) and companion searches (Reid, this volume).

3 Activity

The 2MASS samples allow an initial reconnaissance of chromospheric activity in M and L dwarfs. Two critical questions may be addressed: How *frequent* is chromospheric activity in cool dwarfs? How *strong* is chromospheric activity?

By adding the spectroscopic survey of nearby stars by Hawley et al. [11], the frequency of activity from K7 ($T_{eff} \approx 4200$K down to L8 ($T_{eff} \approx 1400$K?) is plotted in Fig. 6. Activity is defined by Hα in emission with equivalent width > 1Å. Two trends are evident. Over the (stellar) range K7 to M7, the frequency of activity increases as cooler, lower-mass stars are considered. The kinematics of field M dwarfs [11] and observations of open clusters [12] both indicate that the active stars are younger than the inactive stars, and that the increase in the frequency of emission reflects the longer lifetime of high activity levels.

The new result by Gizis et al.[10] is to add the cooler (M7-L8) dwarfs, allowing dwarfs at and below the hydrogen-burning limit to be examined. This reveals (Fig. 6) that the frequency of emission declines rapidly from spectral type M8 down to L4 — with cooler dwarfs in these surveys all lacking chromospheric

74 J.E. Gizis

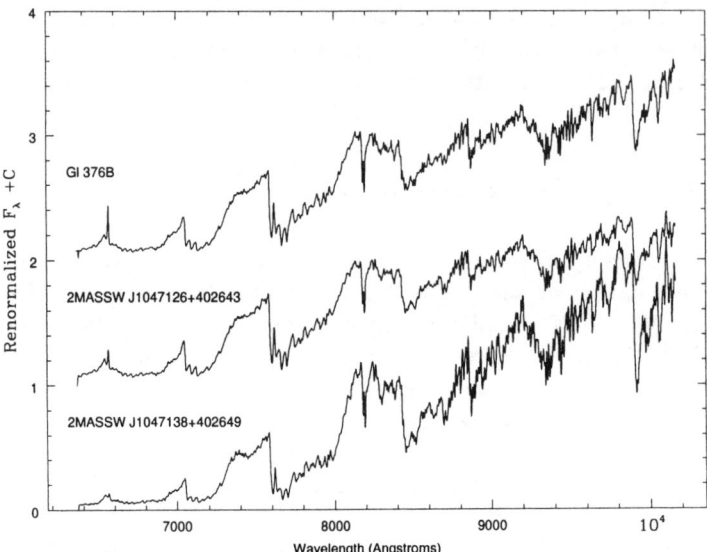

Fig. 3. Spectra of late-M dwarfs illustrating the richness of molecular features and the Hα emission line at 6563Å. The M8 dwarf 2MASSW J1047138+402629 (LP 213-68) has much weaker activity than its M6.5 primary 2MASSW J1047126+402643 (LP 213-67) [9].

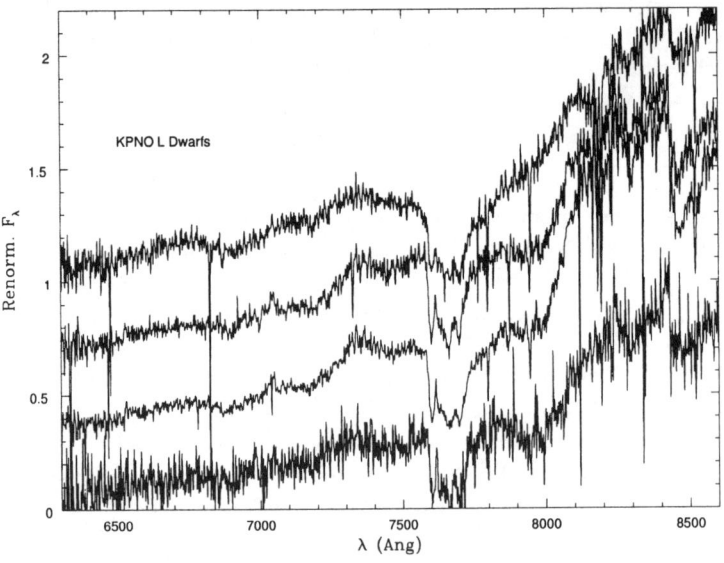

Fig. 4. KPNO 4-meter spectra of 4 nearby L dwarfs with $K_s < 12$. Even in modest exposure times (< 30 minutes), these L dwarfs are accessible to 4-meter telescopes.

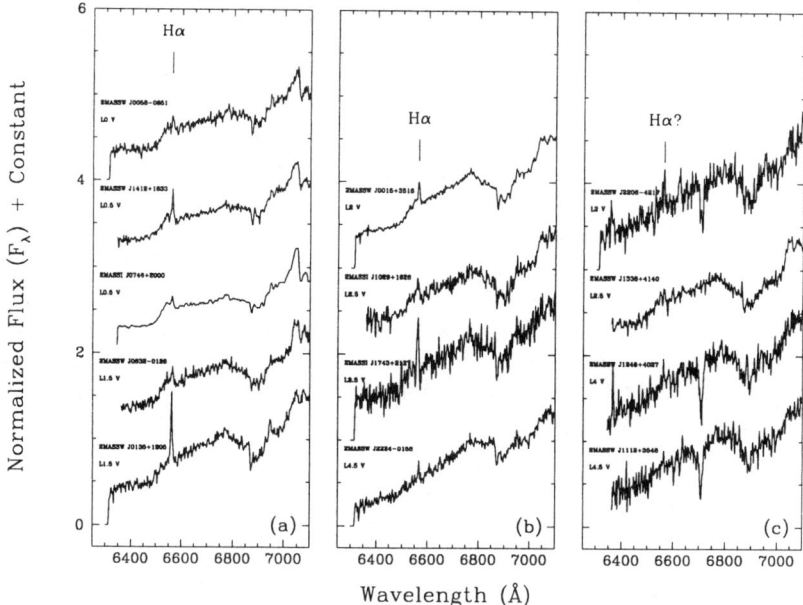

Fig. 5. Keck spectra of L dwarfs from Kirkpatrick et al.[14]. Low (9Å) resolution spectra allows both the Hα (6563Å) emission line and lithium (6708Å) absorption line to be observed. L dwarfs with lithium, although young, generally lack Hα emission.

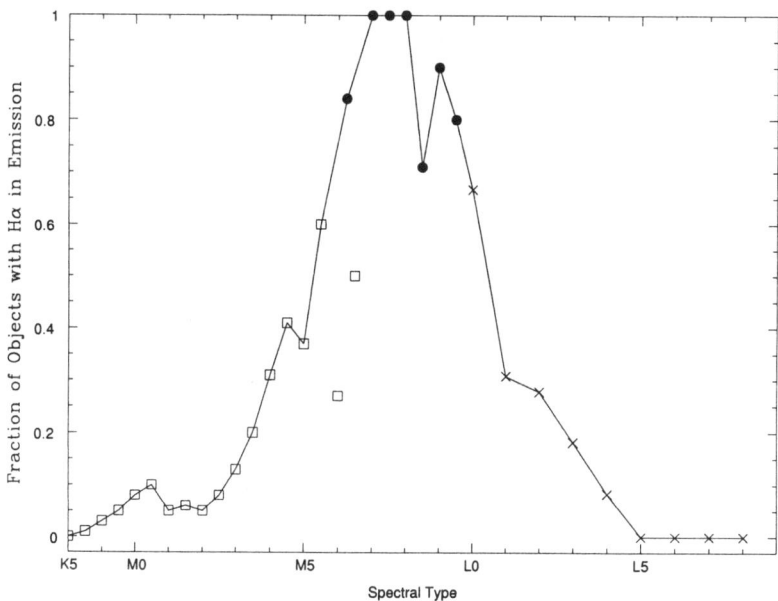

Fig. 6. The frequency of Hα emission for nearby field M and L dwarfs[10,11,14].

activity. As is evident in Fig. 1, over the range M8 to L4 an increasing fraction of objects will be young brown dwarfs rather than old stars.

The mere presence or absence of detectable Hα emission already reveals interesting behavior, but the declining photospheric emission near 6500Å means that Hα is increasingly easy to detect and equivalent width is a poor measure of Hα's importance. We therefore plot the *strength* of Hα emission by considering the ratio of the Hα luminosity to the bolometric luminosity in Fig. 7. The horizontal dotted line indicates the level at which Hawley et al.[11] could detect activity in any M dwarf. It is clear that beyond spectral type M7, even the most active dwarfs would not be considered very strong by the standards of early-to-mid M dwarfs. Furthermore, there is a striking temperature dependence, with the upper envelope of observed activity levels falling by two orders of magnitude over the range M7 to L4. This suggests that effective temperature is the dominant parameter in controlling the activity levels of these cool dwarfs.

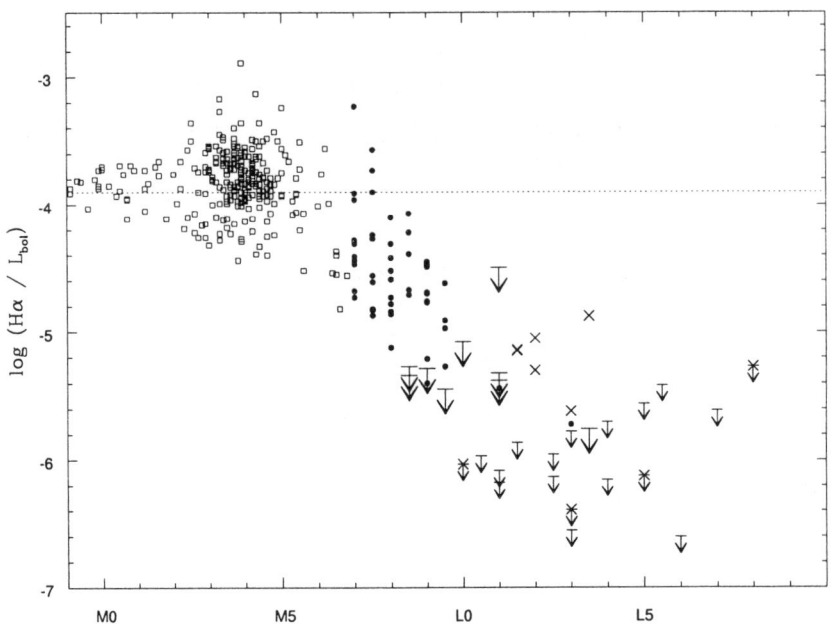

Fig. 7. Activity levels as a function of spectral type for M and L dwarfs. The dotted line at -3.9 is the level at which any M dwarf would be observed in emission. None of the M8 or later dwarfs have activity levels above the -3.9 level, even though such activity is common in mid-M dwarf stars.

4 Adding Age Information

Kinematics provide age information for the M and L dwarf sample. Unlike activity-age relations, where brown dwarfs might not obey the same relations as stars, the "heating" of the brown dwarf velocity distribution should proceed in the same way as for stars (see [4] for the theory of this process). In particular, young brown dwarfs will tend to have low velocities. The disadvantage of kinematics as an age indicator is that it is only statistical — an old star can have a low random space velocity, with the most notable example being the Sun itself!

The tangential velocities of M8 and M9 dwarfs are compared to their chromospheric activity in Fig. 8. It is striking that the expected stellar age-activity relation is not followed. The population of low velocity ($v_{tan} < 20$ km/s), hence younger, M8 and M9 dwarfs have much *weaker* emission than the older population with $v_{tan} > 20$ km/s. One obvious explanation would be that the activity levels of the coolest M dwarfs *increase* with age. However, the theoretical calculations shown in Fig. 1 suggest a different explanation. The young population (corresponding to ages $<\sim 1$ Gyr) are lower-mass stars and brown dwarfs still in their initial cooling phase, while the older, more active population must be hydrogen-burning stars which have largely stabilized at their main-sequence temperature.

For L dwarfs, there is another age and mass diagnostic available in the form of the lithium line[17]. Objects below $\sim 0.06 M_\odot$ are never hot enough to destroy lithium and therefore will show a lithium absorption feature, while more massive stars and brown dwarfs will burn up their lithium in nuclear reactions and therefore lack the feature. At a given temperature (spectral type), L dwarfs with lithium are thus lower-mass and younger than L dwarfs without it (see Fig. 1). Kirkpatrick et al.'s [13,14] data include both lithium and Hα measurements. Consider the L1-L4.5 dwarfs, where lithium is detectable even at low resolution. Only one early-L dwarf, Kelu 1, shows both Hα emission and lithium absorption. Eleven other such early-L dwarfs show Hα emission but do not have lithium absorption. Twelve early-L dwarfs show lithium absorption but do not have Hα emission (four of these have marginal H detections or noise consistent with emission of less than 2). While many L1-L4.5 dwarfs have neither Hα emission nor lithium absorption, it seems clear that the chromospherically active L dwarfs are drawn from an older, more massive population than the lithium L dwarfs. Beyond L4.5, there are no definite cases of Hα emission, although lithium absorption is present for 50% of the L dwarfs. Overall, the L dwarfs show the same properties as the coolest M dwarfs — the younger, less massive dwarfs are *less* active than the older, more massive dwarfs, the opposite of the stellar age-activity relation.

These samples, then, lead to a new view of activity near the hydrogen-burning limit. The dominant parameter is effective temperature, with cooler dwarfs, whether stars or brown dwarfs, only being able to maintain lower activity levels. The mass is a secondary parameter, with lower-mass objects have weaker activity at a given temperature, at least for ages typical of field objects.

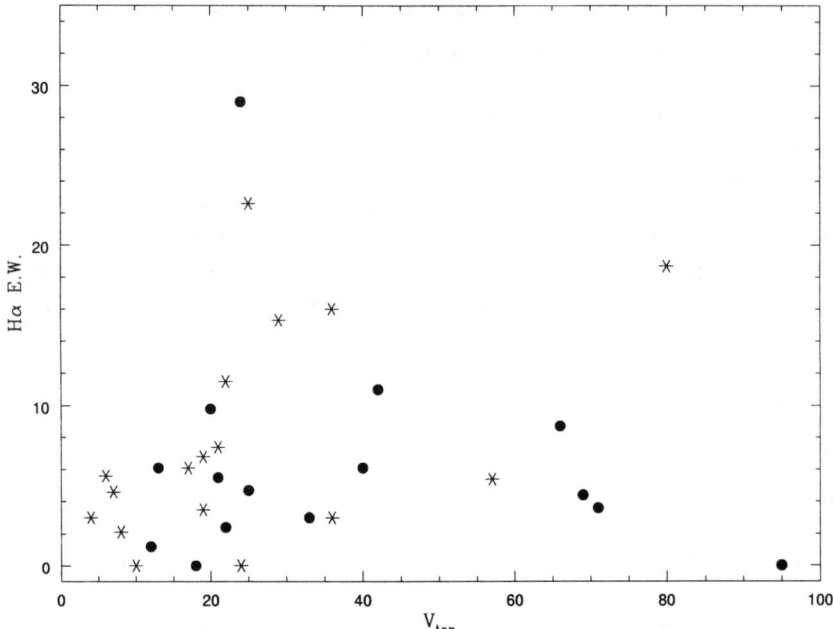

Fig. 8. Emission strength and tangential velocities for M8.0/M8.5 dwarfs (six-pointed stars) and M9.0/M9.5 dwarfs (solid circles). The low-velocity, inactive M dwarfs are a younger population than the high-velocity, active M dwarfs.

The importance of age, rotation rate, and other parameters needs further investigation.

This new view based on 2MASS field dwarfs can be supplemented with observations of cluster brown dwarfs of known age. In Fig. 9, we plot the activity levels for Pleiades brown dwarfs [18–20,25] of age ~ 100 Myr with six-pointed stars. Extremely young brown dwarfs from the ρOph [16] and the σOri [3,26] are also shown. For comparison, the (old) field dwarfs with $v_{tan} > 20$ km/s are plotted as solid circles and the (young) field dwarfs with $v_{tan} < 20$ km/s are plotted as open circles.

Despite their youth, activity in the Pleiades brown dwarfs obey the same fall-off with temperature seen in the field dwarfs. Their youth is evidently offset by their lower mass, resulting in unremarkable activity levels. Indeed, no emission is detectable in the Pleiades L dwarf, even though activity is seen in some field L dwarfs of similar spectral type.

The very young (< 10 Myr) brown dwarfs show that the relation between Hα emission and age is not simple. Half of the young cluster brown dwarfs show weak emission, consistent with the field M and L dwarfs, but half show much stronger emission. Like the Pleiades L brown dwarf, the σ Ori L-type brown dwarf has no detectable activity despite its youth. In the case of the ρOph brown dwarf, the

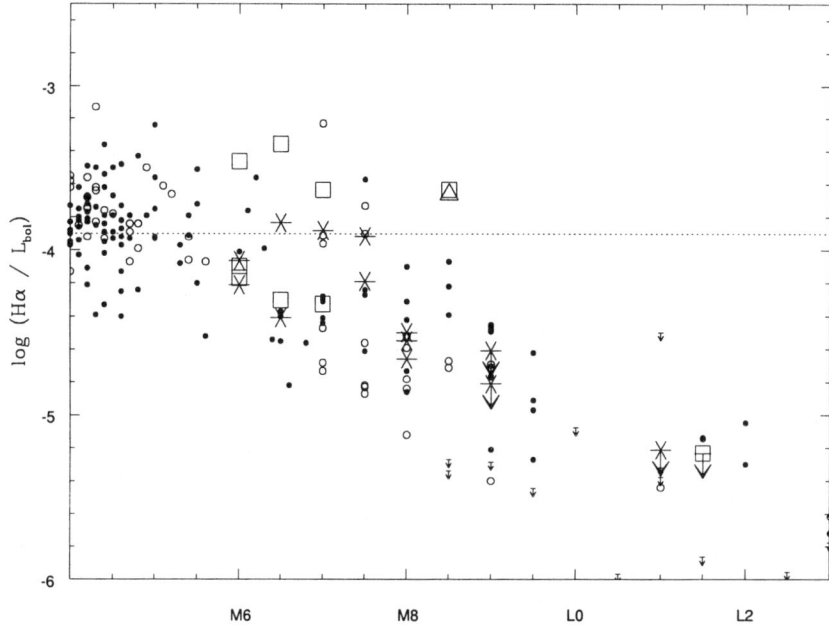

Fig. 9. The Hα luminosity relative to the bolometric luminosity as a function of spectral type for both cluster brown dwarfs and field dwarfs. Brown dwarfs from the σ Ori cluster ($< 10^7$ years), ρ Oph ($< 10^7$ years), Pleiades ($\sim 10^8$ years) are shown as open squares, open triangles, and six-pointed stars respectively. Note that both cluster L dwarfs have only upper limits on the detected Hα emission. The field M dwarfs are plotted as open circles if $v_{tan} < 20$ km/s and solid circles for higher velocities.

mid-IR excess strongly suggests that the emission is due to a disk[16], and Gizis et al.[10] suggest that the 'active' brown dwarfs are actually those with disks, while the relatively inactive brown dwarfs simply show the weak chromospheres expected by analogy with the field dwarfs.

As kinematic information becomes available for more brown dwarfs, it will be possible to test evolutionary models statistically. A first step is shown in Fig. 10, where tangential velocity is plotted as a function of spectral type. The tangential velocities are based on USNO parallaxes and proper motions (Dahn, priv. comm.). Objects with lithium are shown as open circles, those without lithium are shown as solid triangles. The results are reassuring — the lithium L dwarf population is clearly kinematically cooler, hence younger, than the non-lithium L dwarfs, confirming theoretical expectations. A larger sample will enable trends with spectral type, activity, or other properties to be investigated. Addition of radial velocities will allow the full three-dimensional space velocity distribution to be considered.

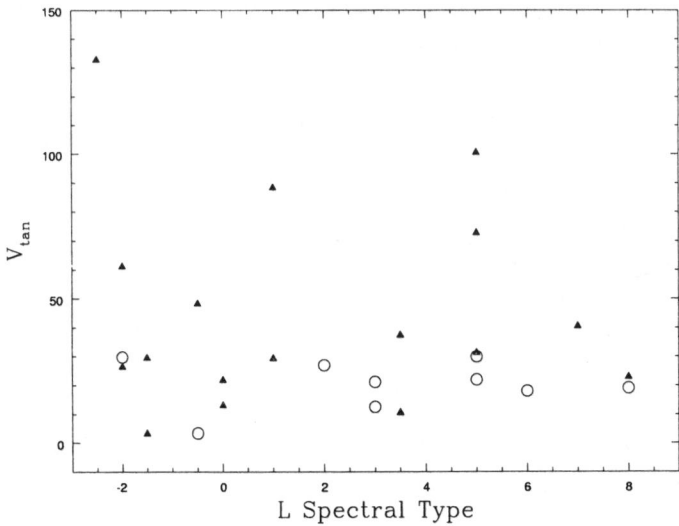

Fig. 10. Tangential velocities from U.S.N.O. astrometry for L dwarfs with (open circles) and without (solid triangles) lithium. The lithium L dwarfs have smaller velocities and hence are a younger propulation.

5 Further Considerations on Magnetic Activity

Theoretical investigations are needed to understand the temperature and mass dependence of Hα emission from the chromospheres of L dwarfs. Does the lack of emission mean that there is no chromosphere? The latest X-ray observations of an M8 star [7] and M9 brown dwarf [23] indicate that quiscent X-ray coronae are not detectable. The disappearance of coronae is consistent with the observed weakening of the chromosphere. Clearly X-ray observations of L dwarfs would be of great interest to confirm this trend.

Although the quiscent activity is already disappearing at M8-M9 spectral types, it is clear that magnetic fields have not entirely disappeared. In Gizis et al.'s [10] spectroscopic survey, 7% of the late-M dwarfs were caught in a flare event. Liebert et al.[15] have reported a huge flare in a 2MASS M9.5 dwarf. Reid et al.[22] observed a flare in the well-known M9.5 dwarf BRI0021, the prototype 'inactive' rapid-rotator [2], and also estimate a 7% flare rate. Flaring activity apparently occurs in objects both with and without quiscent chromospheric emission. X-ray flares have also been observed in the M9 brown dwarf LP944-20 [23] with a similar estimated flare rate. Despite the weakening of both the chromosphere and corona, flaring is frequent and strong in the coolest M dwarfs, suggesting that magnetic fields persist. A search for flaring activity among both active and inactive L dwarfs will be of interest, and should naturally occur as 'standard' L dwarfs are reobserved. The photometric monitoring reported by

Bailer-Jones (this volume) should show whether or not 'starspots' exist in these dwarfs.

Two exceptions to the general trend are known. The M9.5 dwarf PC0025 was discovered due to its strong (EW$\approx$ 300Å) Hα emission which has persisted for a decade[24]. Is this a case of a strong chromosphere, or is the emission due to something else? It must be a rare or short-lived phenomenon, since no other field M dwarfs like it have been discovered. More recently, a 2MASS T dwarf has been discovered which has strong, persistant Hα emission [5]. No other T dwarfs, or even late-L dwarfs, have such emission. Is this a chromosphere on a 1000K object, and if so, why is it unique? Other scenarios for the emission are possible[5].

6 Conclusions

The kinematics of brown dwarfs offer a powerful way to measure ages. Tangential velocities are relatively easy to obtain and should allow the relative ages of M, L, and T dwarf samples to be measured. Preliminary work has already confirmed that lithium L dwarfs are younger than other L dwarfs. Age information should help constrain models of the field brown dwarf mass function which are sensitive to the Galactic star formation history [21].

For chromospheric activity, there is a good news and bad news. The bad news is that activity is not a very good age indicator. In particular, Hα emission is more likely to be a sign of age in field late-M and L dwarfs than youth, the opposite of the traditional interpretation. The good news is that strong trends are seen in the observations, raising interesting questions. Why do chromospheres — and coronae — disappear at the these temperatures? Why is there a mass dependence? Why does flaring persist? What are the unusual objects with strong, persistent emission? The observations beg for theoretical investigations. As new objects are discovered and new observations become possible, we should expect further surprises.

With the completion of 2MASS and further follow-up, new tests of the scenario in this paper will be made. 2MASS has proven to be sensitive to wide L dwarf companions of main sequence stars. Since ages can be independently estimated from the primary star (Kirkpatrick, this volume), tests of the activity-age relation will be possible. Additional cluster brown dwarfs will also be of great help. 2MASS is sensitive to L dwarfs in the Hyades — while an initial search of a portion of the cluster suggests that the Hyades has lost most of its brown dwarfs[8], it may be hoped that at least one L dwarf remains for 2MASS to discover.

Acknowledgments

I would like to thank the 2MASS Rare Objects Team (Jim Liebert, Davy Kirkpatrick, Neill Reid, Dave Monet, Conard Dahn, Adam Burgasser), Mike Skrutskie, and Suzanne Hawley for many discussions of stars and brown dwarfs. I acknowledge the support of the Jet Propulsion Laboratory, California Institute of Technology, which is operated under contract with NASA. This publication makes use of data products from 2MASS, which is a joint project of the University of Massachusetts and IPAC/Caltech, funded by NASA and NSF. I was a Visiting Astronomer, Kitt Peak National Observatory, National Optical Astronomy Observatories, which is operated by the Association of Universities for Research in Astronomy, Inc. (AURA) under cooperative agreement with the NSF.

References

1. I. Baraffe, G. Chabrier, F. Allard, & P.H. Hauschildt, A&A, **337**, 403 (1998)
2. G. Basri, & G.W. Marcy, AJ, **109**, 762 (1995)
3. V.J.S. Béjar, M.R. Zapetero Osorio, & R. Rebolo, ApJ, **521**, 671 (1999)
4. J. Binney & S. Tremaine, Galactic Dynamics, Princeton University Press (1987)
5. A.J. Burgasser, J.D. Kirkpatrick, I.N. Reid, J. Liebert, J.E. Gizis, & M.E. Brown, AJ, **120**, 473 (2000)
6. A. Burrows et al., ApJ, **491**, 856 (1997)
7. T.A. Fleming, M.S. Giampapa, & J.H.M.M. Schmitt, ApJ, **533**, 372 (2000)
8. J.E. Gizis, I.N. Reid, & D.G. Monet, AJ, **118**, 997 (1999)
9. J.E. Gizis, D.G. Monet, I.N. Reid, J.D. Kirkpatrick, & A.J. Burgasser, MNRAS, **311**, 385 (2000)
10. J.E. Gizis, D.G. Monet, I.N. Reid, J.D. Kirkpatrick, J. Liebert, & R.J.Williams, AJ, **120**, 1085 (2000)
11. S.L. Hawley, J.E. Gizis, & I.N. Reid, AJ, **112**, 2799 (1996)
12. S.L. Hawley, I.N. Reid, J.E. Gizis,& P.B. Byrne, 'Chromospheric Activity in Low Mass Stars: Observational Results from Clusters and the Field'. In: Solar and Stellar Activity: Similarities and Differences, ASP Conference Series **158**, ed. C. J. Butler & J. G. Doyle., p.63 (1999)
13. J.D. Kirkpatrick, I.N. Reid, J. Liebert, R.M. Cutri, B. Nelson, C.A. Beichman, C.C. Dahn, D.G. Monet, J.E. Gizis, & M.F. Skrutskie, ApJ, **519**, 802 (1999)
14. J.D. Kirkpatrick, I.N. Reid, J.E. Gizis, A.J. Burgasser, J. Liebert, D.G. Monet, C.C. Dahn, & B. Nelson, AJ, **120**, 447 (2000)
15. J. Liebert, J.D. Kirkpatrick, I.N. Reid, & M.D. Fisher, ApJ, **519**, 345 (1999)
16. K.L. Luhman, J. Liebert, & G.H. Rieke, ApJ, **489**, L165 (1997)
17. A. Maguzzù, E.L. Martiín, & R. Rebolo, ApJ, **404**, L17 (1993)
18. E.L. Martín, R. Rebolo, & M.R. Zapatero-Osorio, ApJ, **469**, 706 (1996)
19. E.L. Martín, G. Basri, J.E. Gallegos, R. Rebolo, M.R. Zapatero Osorio, ApJ, **499**, L61 (1998)
20. E.L. Martín, G. Basri, M.R. Zapatero Osorio, R. Rebolo, & R.J. Garcia Lopez, ApJ, **499**, L41 (1998)
21. I.N. Reid, J.D. Kirkpatrick, J. Liebert, A. Burrows, J.E. Gizis, A. Burgasser, C.C. Dahn, D. Monet, R. Cutri, C.A. Beichman, & M. Skrutskie, ApJ, **521**, 613 (1999)
22. I.N. Reid, J.D. Kirkpatrick, J.E. Gizis, & J. Liebert, ApJ, **527**, L105 (1999)
23. R.E. Rutledge, G. Basri, E.L. Martín, & L. Bildsten, ApJ, **538**, L141 (2000)
24. D.P. Schneider, J.L. Greenstein, M. Schmidt, & J.E. Gunn, AJ, **102**, 1180 (1991)
25. M.R. Zapatero Osorio, R. Rebolo, E.L. Martín, G. Basri, A. Maguzzu, S.T. Hodgkin, R.F. Jameson, & M.R. Cossburn, ApJ, **491**, L81 (1997)
26. M.R. Zapatero Osorio, V.J.S. Bejar, R. Rebolo, E.L. Martín, & G. Basri, ApJ, **524**, L115 (1999)

Infrared Spectroscopy of Brown Dwarfs: the Onset of CH$_4$ Absorption in L Dwarfs and the L/T Transition

T.R. Geballe[1], K.S. Noll[2], S.K. Leggett[3], G.R. Knapp[4], and X. Fan[4], and D. Golimowski[5]

[1] Gemini Observatory, 670 N. A'ohoku Pl., Hilo, HI 96720, USA
[2] Space Telescope Science Institute, 3700 San Martin Dr., Baltimore, MD 21218, USA
[3] Joint Astronomy Centre, 660 N. A'ohoku Pl., Hilo, HI 96720, USA
[4] Princeton University Observatory, Princeton, NJ 08544, USA
[5] Dept. of Physics and Astronomy, Johns Hopkins Univ., 3701 San Martin Dr., Baltimore, MD 21218, USA

Abstract. We present infrared spectra of brown dwarfs with spectral types from mid-L to T. The 0.9-2.5 μm spectra of three dwarfs found by the Sloan Digital Sky Survey contain absorption bands of both methane and carbon monoxide and bridge the gap between late L and previously observed T dwarfs. These dwarfs form a clear spectral sequence, with CH$_4$ absorption increasing as the CO absorption decreases. Water vapor band strengths increase in parallel with the methane bands and thus also link the L and T types. We suggest that objects with detectable CO and CH$_4$ in the H and K bands should define the earliest T subclasses. From observations of bright (K $\leq$ 13 mag) L dwarfs found by 2MASS, we find that the onset of detectable amounts of CH$_4$ occurs near spectral type L5. For this spectral type methane is observable in the 3.3 μm ν_3 band only, and not in the overtone and combination bands at H and K.

1 Introduction

The infrared spectra of T type brown dwarfs are profoundly affected by the molecules methane (CH$_4$) and water (H$_2$O). Huge bites are taken out of the spectra by absorption bands of these molecules. At other infrared wavelengths, adjacent to some of these absorption bands, the atmospheres of the dwarfs are remarkably transparent. A photosphere with an effective temperature of 950 K, the value for Gliese 229B, might naively be expected to emit its maximum flux density near the peak of a blackbody of that temperature, i.e., just longward of 3 μm. However, as is shown in Fig. 1 (upper panel) the maximum for Gl 229B occurs in the J band, near 1.2 μm, almost a factor of three shorter wavelength. The strongest band of CH$_4$ is in fact centered at 3.3 μm (very close to the peak of a 950 K black body) and virtually no radiation emerges from the brown dwarf near that wavelength. In the J, H, and K bands the locations of the CH$_4$ and H$_2$O absorptions give a T dwarf roughly the JHK colours of a very hot star.

The infrared spectrum of an L dwarf (Fig. 1, lower panel) also is badly eaten away by molecular absorptions, many of which are different than those affecting a T dwarf. However, the maximum flux density occurs close to 1.5 μm, the peak

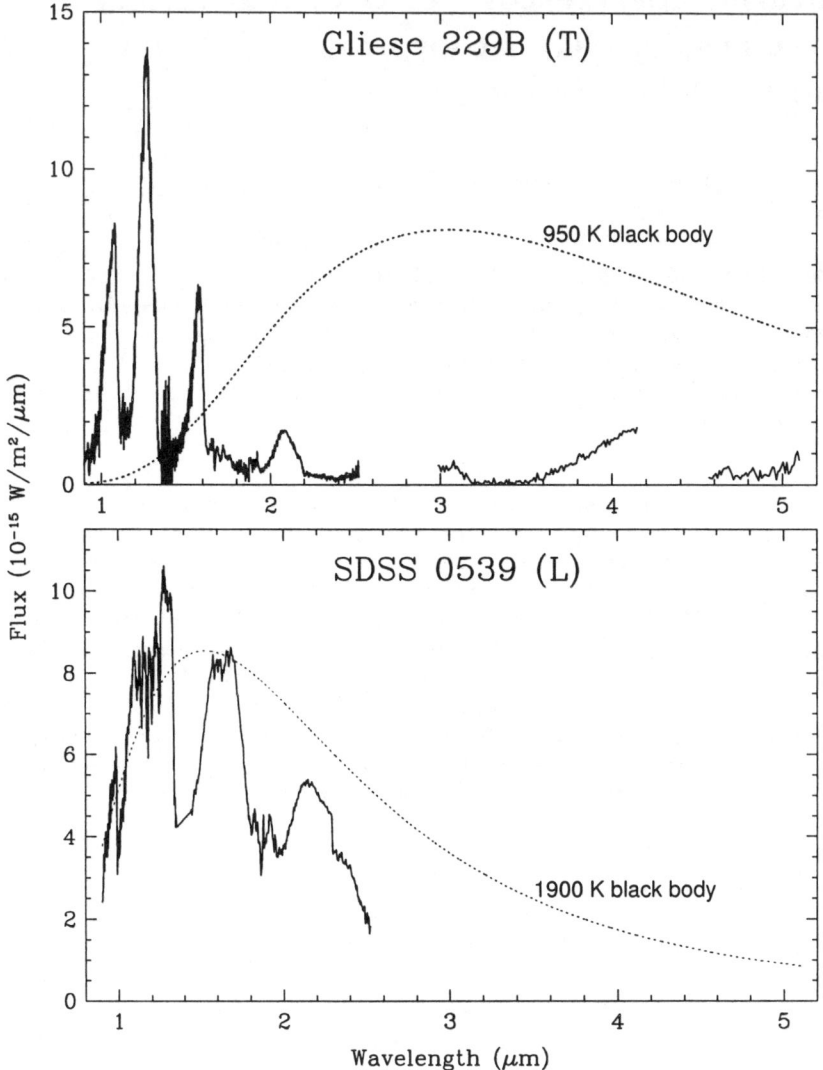

Fig. 1. Spectra of representative T and L type brown dwarfs, Gl 229B [5],[13],[12] and SDSS 0539 [9] compared with black body functions (of arbitrary strengths) corresponding to the effective temperatures of the brown dwarfs. The original Gl 229B data have been recalibrated [6], [10].

of a blackbody whose temperature is ~2000 K, roughly the effective temperature of the dwarf. In addition, the JHK colours of L dwarfs are much closer to what one naively would expect for an object of that temperature.

Thus in the key J, H, and K bands the spectral transition of a brown dwarf from L to T is a huge change, much greater than between any other two neighboring spectral types. The transformation is largely due to the increasing stability of methane at lower temperatures, its increasing abundance as the overwhelming presence of hydrogen drives the chemical equilibrium of carbon-bearing species toward dominance by CH_4, and the resulting onset of strong absorption by CH_4 in a variety of combination and overtone bands in the 1-2.5 μm region.

The bands of CH_4 in the 1-2.5μm interval are strong in T dwarfs, but do not correspond to fundamental vibrations of the molecule. Those much stronger bands occur at longer wavelengths where, because of poor atmospheric transmission and high sky and telescope backgrounds, few spectra of L and T dwarfs have been obtained.

2 T-Dwarf Spectra Before 2000

Following the discovery of Gl 229B, it took five years before the Sloan Digital Sky Survey (SDSS), the Two-Micron All Sky Survey (2MASS), and the European Southern Observatory (ESO) found additional T dwarfs [14],[2],[4],[15]. The 1-2.5 μm spectra of the objects they discovered are nearly identical to that of Gl 229B. There are in fact some differences between them, probably related to differences in temperature and surface gravity, but they are fairly subtle. At the end of 1999 no objects were known with spectra located in the huge gap, demonstrated in Fig. 1, between the latest L type, which showed no CH_4 absorption at 1-2.5 μm and the T types, in which very strong CH_4 bands are present. It was not clear at the time whether this lack was an observational selection effect or indicated that the transition from latest L-types to then known T-types was rapid compared to the overall cooling time of an L dwarf, encompassing only a relatively narrow temperature range.

It now is evident that the lack of objects in the transition region was largely observational selection. This is demonstrated in Fig. 2, in which $J - K$ is plotted vs. $i - z$ for L and T dwarfs. The area at the top left of the figure is the domain of the L dwarfs, the area at bottom right is the domain of the classical T dwarfs. In 1999 the majority (5) of all (8) published T dwarf identifications were from 2MASS, which surveys the sky at J, H, and K *only*. In Fig. 2 it can be seen that the T dwarfs have blue or bluer $J - K$ colours than all but the hottest stars. Objects with such colours form a very small subset of the 2MASS catalogue. Determining if such objects are candidate brown dwarfs is relatively straightforward, requiring comparison with the Palomar Sky Survey. However, if, as expected, during the transition between L and T, brown dwarfs follow a direct path in Fig. 2 between these two domains, their $J - K$ colours will pass through the same range of values as the myriads of cool main sequence stars and they will be photometrically indistinguishable from such stars by 2MASS.

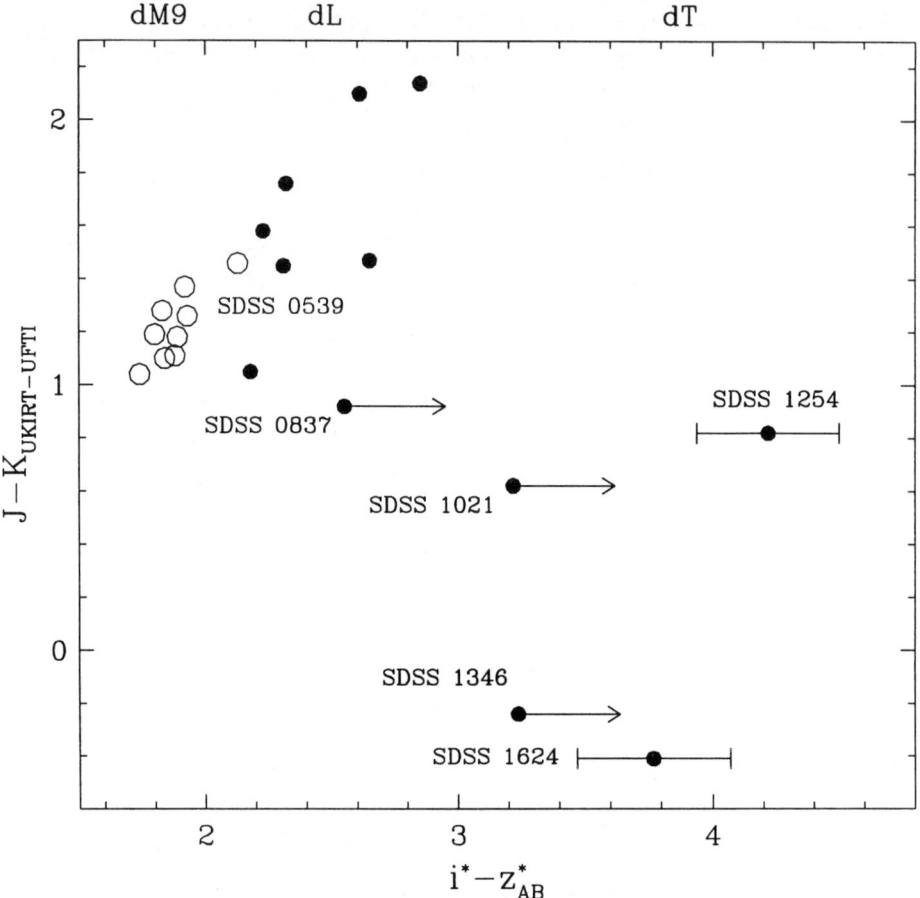

Fig. 2. UKIRT(UFTI) J-K vs. SDSS $i - z$ [9]. The objects in the top left quadrant are late M and L dwarfs; those at the bottom are classical T dwarfs.

In contrast, the transition objects would be expected to develop even redder and more unusual $i - z$ colours and, as in the case of T dwarfs, to be readily singled out by SDSS. However, by the end of 1999 the two objects at the bottom of Fig. 2 were the only SDSS objects known to be later than L.

3 Transition Objects

In early 2000 three new objects were identified as candidate T dwarfs by SDSS, on the basis of their $i - z$ colours. JHK photometry at UKIRT [9] showed that these objects (SDSS 0837, SDSS 1021, and SDSS 1254) inhabited the region in Fig. 2 between the previously known L and T dwarfs. Spectra of them were obtained at the United Kingdom Infrared Telescope (UKIRT) in late February and March, 2000 and are displayed in Fig. 3, together with those of a mid-L

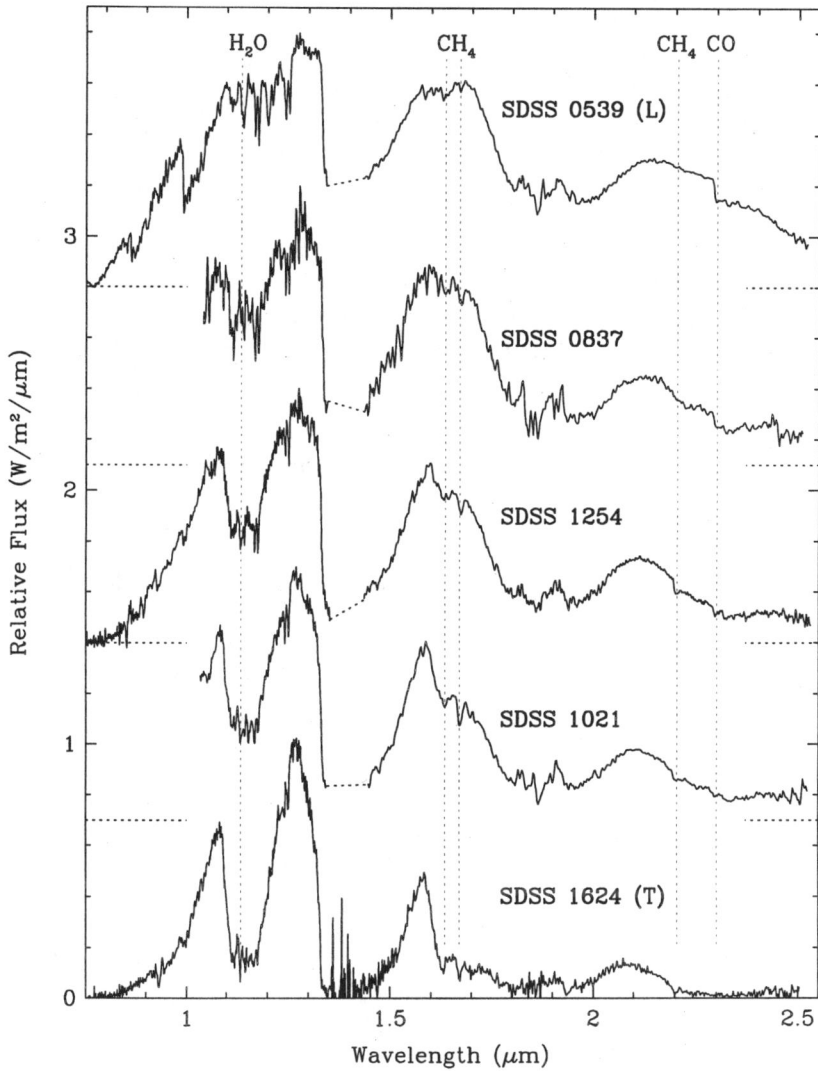

Fig. 3. Spectra of three transition objects together with a mid-L dwarf (top) and a classical T dwarf (bottom) [9]. The dashed vertical lines mark the wavelengths of bands (or, in the case of CO, the 2-0 band head) whose strengths change rapidly in the transition region. The horizontal dashed lines are zero flux density levels for the individual spectra.

dwarf and a previously known SDSS T dwarf. These three spectra fortuitously delineate the transition between L and T fairly evenly, with CO overtone bands at 2.3-2.4 μm gradually disappearing from top to bottom in the figure and the methane bands at 1.6-1.8 μm and 2.2-2.5 μm gradually strengthening. One of the largest spectral changes is in the strength of the water band ($\nu_1 + \nu_2 + \nu_3$) at 1.15 μm. That band is weak in the L dwarf at the top of Fig. 3, but increases steadily in the transition objects, and is nearly totally absorbing in the classical T dwarf.

At the time of this conference, SDSS has discovered comparable numbers of classical T dwarfs and transition objects. Although the numbers of each are few, they suggest that both transition dwarfs and classical T dwarfs occupy substantial temperature ranges. Hence it is likely that many more of each kind will be found by SDSS.

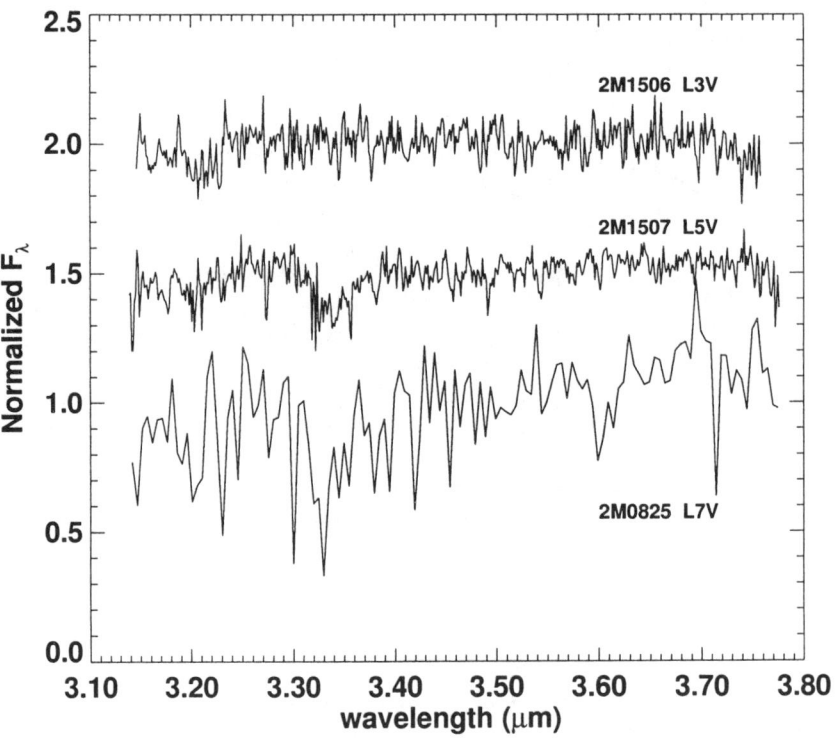

Fig. 4. Spectra of three L-type brown dwarfs in the region of the methane ν_3 fundamental band [11]. The broad absorption present in the two later type dwarfs is the methane Q-branch.

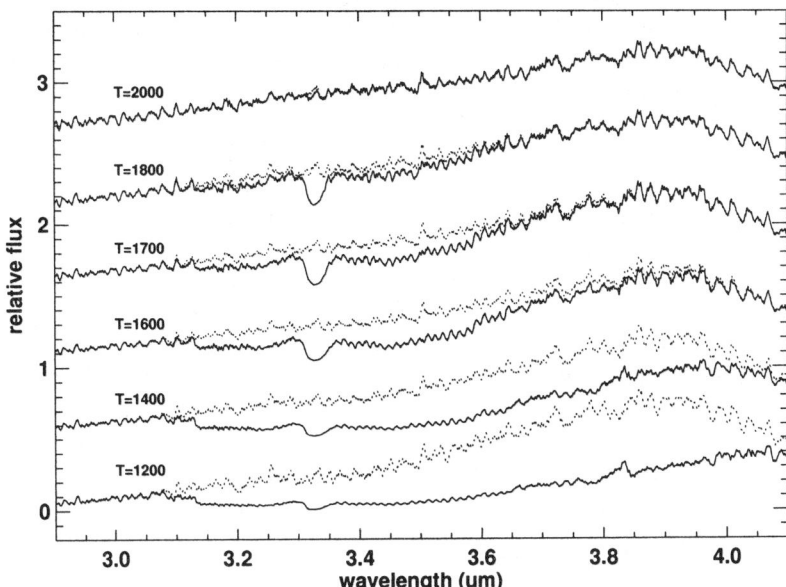

Fig. 5. Model 2.9-4.1 μm spectra for dwarfs with effective temperatures ranging from 1200 K to 2000 K. Dashed curves contain no methane, and the spectral features are primarily due to water vapor. Solid curves have absorption from both H_2O and CH_4. The ν_3 Q branch first becomes detectable at $T_{eff} \approx 1900$ K. The P and R branches begin to significantly suppress the surrounding H_2O pseudo-continuum by $T_{eff} \approx 1700$ K.

4 Methane Absorption in L Dwarfs

As mentioned earlier, the most intense bands of methane occur outside the JHK region, at longer wavelengths. The strongest of these is the ν_3 band centered at 3.3 μm. This band, one-hundred or more times stronger than the any of the CH_4 bands in the 1-2.5 μm interval, ought to appear at higher temperatures than the 1-2.5 μm bands and must already be very strong in the transition objects discussed above. Thus, one would predict that in a late L dwarf the ν_3 band should be detectable, as perhaps the first sign of the change taking place in its atmospheric chemistry.

In May 2000 we used UKIRT and CGS4 to search for this band in a set of L dwarfs of different spectral types. As shown in Fig. 4 a broad absorption feature, which we identify as the Q branch of the ν_3 band of CH_4, was detected in the two latest objects observed, 2MASS 1507 and 2MASS 0825 [11], classified as L5 and L7.5, respectively [7]. The observed feature is slightly shifted with respect to the strong Q branch absorption in the earth's atmosphere, due to the higher temperatures in the dwarfs, which cause the excitation of higher J levels of the molecule. The extension of the absorption to longer wavelengths makes the Q branch more readily detectable from the ground. We also obtained a spectrum of 2MASS 1507 in the K band, finding no evidence for methane absorption in that band. No K band spectrum of the later object, 2MASS 0825 was obtained.

Models of L dwarf spectra in the 3-4 μm region (Fig. 5) show the growth of methane absorption with decreasing effective temperature. Methane abundances at each layer of the model atmosphere were calculated using published chemical equilibrium abundance profiles [3]. The strength of the methane absorption is weakly dependent on model assumptions about cloud structure; the models in Fig. 5 are for a clear atmosphere. The weakness of the observed methane Q branches in the spectra of 2MASS 1507 and 2MASS 0825 (Fig. 4) and the absence of absorption due to the P and R branches suggest that these dwarfs have effective temperatures near 1800 K. This is in general agreement with temperatures suggested in [1], but higher than the temperature inferred in [8] for these spectral classes. Because of methane's temperature-sensitive chemistry and because methane is expected to be found well above any condensation clouds, the ν_3 band promises to be an effective indicator of temperature for late L dwarfs.

5 Classification Issues

The discovery of transition objects and the detection of absorption by methane in L-type dwarfs bring to the fore several issues pertaining to classification and terminology.

1. *Do the three "transition objects" belong to type L or T?* Our view is that these objects should define the earliest part of the T sequence, which up to now shows very little spectral variation compared to that of type L. Moreover, the latter class appears to be running short of available subtypes in the classification systems proposed at present.

2. *However they are classified, what spectral features should be used in defining the transition sub-types?* As is shown in this paper, large changes between the spectra of late L-type and classical T-type objects occur in the 1-2.5 μm region. These changes are much greater than those that occur at shorter wavelengths for the same objects. Furthermore, because the optical–near infrared colours of "transition" dwarfs and classical T dwarfs both are extremely red, they are as or more tractable to observe in the J, H, and K bands than at optical wavelengths, Many of them are easily observable in these bands by 4 meter class telescopes. These characteristics make 1-2.5 μm the most attractive interval for defining the classification scheme. Because of their strong variation with temperature we regard the methane bands at 1.6-1.7 μm and 2.2-2.4 μm and the water band at 1.15 μm (see Fig. 3) as promising candidates to employ in defining sub-types. Although the methane bands at 2.2 μm and longer are stronger that the bands at 1.6-1.7 μm, contamination of the former by CO and the variable depression of the continuum peak near 2.1 μm by the H_2 pressure-induced dipole absorption (which is more pressure sensitive than temperature sensitive) may make use of the K band problematic.

3. *Should the discovery of CH_4 at 3.3 µm in mid and late L types affect the proposed L and incipient T classification schemes?* We believe not. In large part the reasons have to do with practicality. Accurate ground-based measurements in the L band are considerably more difficult to obtain than similar measurements at shorter wavelengths. In the L band obtaining signal-to-noise ratios sufficiently high for purposes of classification will be possible for only a small fraction of all L-type dwarfs, compared to the much larger fractions that will be observable spectroscopically at optical and shorter infrared wavelengths. Thus we suggest that dwarfs showing the ν_3 and (possibly) longer wavelength fundamental bands of CH_4, but not the combination bands in the 1-2.5 µm region should remain as L-types.

Although we have suggested that the 3 µm methane band not enter into classification schemes, we point out that until a more comprehensive set of JHK-band spectra are in hand for late L (in currently proposed schemes) and "transition" dwarfs, it probably is premature to define the L-T boundary. Finally, the detection of methane in L-type objects indicates that the term "methane dwarf," which previously has been used interchangeably with "T dwarf," should now be used with caution.

References

1. G. Basri et al.: ApJ **538**, 363 (2000)
2. A.J. Burgasser et al.: ApJ **522**, L65 (1999)
3. A. Burrows, C.M.Sharp: ApJ **512**, 843 (1999)
4. J.G. Cuby, P. Saracco, et al.: A&A **349**, L41 (1999)
5. T.R. Geballe, S.R. Kulkarni, C.E. Woodward, G.C. Sloan: ApJ **467**, L101 (1996)
6. D.A. Golimowski, C.J. Burrows, S.R. Kulkarni, B.R. Oppenheimer, R.A. Brukardt: AJ **115**, 2579 (1998)
7. J.D. Kirkpatrick et al.: AJ, 120, 447 (2000)
8. J.D. Kirkpatrick et al.: ApJ, 519, 802 (1999)
9. S.K. Leggett, T.R. Geballe et al.: ApJ **536**, L35 (2000)
10. S.K. Leggett, D.W. Toomey, T.R. Geballe, R.H. Brown: ApJ **517**, L139 (1999)
11. K.S. Noll, T.R. Geballe, S.K. Leggett, M.S. Marley: ApJ, **541**, L75 (2000)
12. K.S. Noll, T.R. Geballe, M.S. Marley: ApJ **489**, L87 (1997)
13. B.R. Oppenheimer, S.R. Kulkarni, K. Matthews, M.H. Van Kerkwijk: ApJ **502**, 932 (1998)
14. M.A. Strauss, X. Fan, et al.: ApJ **522**, L61 (1999)
15. Z.I. Tsvetanov et al.: ApJ **531**, L61 (2000)

Surface Features, Rotation and Atmospheric Variability of Ultra Cool Dwarfs

C.A.L. Bailer-Jones

Max-Planck-Institut für Astronomie, Königstuhl 17, D-69117 Heidelberg, Germany

Abstract. Photometric I band light curves of 21 ultra cool M and L dwarfs are presented. Variability with amplitudes of 0.01 to 0.055 magnitudes (RMS) with typical timescales of an hour to several hours are discovered in half of these objects. Periodic variability is discovered in a few cases, but interestingly several variable objects show no significant periods, even though the observations were almost certainly sensitive to the expected rotation periods. It is argued that in these cases the variability is due to the evolution of the surface features on timescales of a few hours. This is supported in the case of 2M1145 for which no common period is found in two separate light curves. It is speculated that these features are photospheric dust clouds, with their evolution possibly driven by rotation and turbulence. An alternative possibility is magnetically-induced surface features. However, chromospheric activity undergoes a sharp decrease between M7 and L1, which may be related to the observed greater occurrence of variability in objects later than M9.

1 Introduction

Large numbers of very cool compact objects – namely low mass stars, brown dwarfs and giant gas planets – have only recently become available through large scale surveys, particularly in the far red and near infrared where these objects radiate most of their energy. This has presented the opportunity to systematically study their intrinsic properties. Indeed, the L dwarf sequence has recently been introduced to account for the increasing number of objects found with effective temperatures apparently in the range 2200–1300 K [4][14][15][21], and the T dwarf class covers the even cooler objects (similar to Gl 229B) now being discovered [6][17][23].

For a number of reasons, these ultra cool dwarfs are likely to have interesting and complex atmospheres. First, they are fully convective. Second, many are rapid rotators [4]. Third, at these low temperatures, solid dust particles, or condensates, form. These first two properties may be a driver for atmospheric dynamics, or weather, and the presence of dust makes it possible that large-scale clouds form. Although these objects cannot (yet) be resolved, the presence of weather patterns can be investigated by via accurate photometric monitoring, which is the subject of this article.

Surface Features, Rotation and Atmospheric Variability of Ultra Cool Dwarfs

1.1 Previous Work

Initial attempts to observe variability in ultra cool dwarfs have met with mixed results. Terndrup et al. [28] searched for rotational modulation of the light curves of eight M-type stars and brown dwarfs in the Pleiades. They derived periodicities for two low mass stars, but found no significant variability in the rest of the sample. Tinney & Tolley [29] found some evidence for variability in an M9 brown dwarf with an amplitude of 0.04 magnitudes over a few hours, but detected no variability above 0.1 magnitudes in an L5 dwarf. Nakajima et al. [22] found variability in the near infrared spectrum of a T dwarf over a period of 80 minutes. In a precursor to the present project, Bailer-Jones & Mundt [1] observed six M and L dwarfs and found variability in one (2M1145), to which a tentative period of 7.1 hours was assigned (pending confirmation).

1.2 Observational Sample

This paper reports results from an observational program to monitor photometric I band variability in a sample of 21 late M and L dwarfs (Table 1). Only about 30 L dwarfs were known at the time of the observations, thus greatly limiting the choice of targets. Objects brighter than $I = 19.0$ were preferentially selected, but there are no other (known) selection biases. Ten of the targets are field dwarfs discovered by the Two Micron All Sky Survey (2MASS) and the Sloan Digital Sky Survey (SDSS). Five objects are members of the Pleiades (age 120 Myr): Teide 1 and Calar 3 are confirmed brown dwarfs, Roque 11 and 12 are probably brown dwarfs, and Roque 16 is very close to the hydrogen burning limit (so its status is uncertain). The six remaining objects are candidate members of the σ Orionis cluster (age 1–5 Myr), with masses between 0.02 and $0.12\,M_\odot$.

2 Observations and Data Reduction

The targets were observed with a CCD camera on the 2.2m telescope at the Calar Alto Observatory (Spain) over three periods: January 1999 (AJD 1187.4–1192.8, hereafter 99-01), September 1999 (AJD 1432.8–1436.2, hereafter 99-09) and February 2000 (AJD 1601.8–1607.2, hereafter 00-02). AJD=JD-2450000. Exposure times of 8 minutes were used for most objects to achieve a signal-to-noise ratio (SNR) for the target objects of at least 100. Objects were observed repeatedly each night for several nights to construct a light curve with typically 30 points. The data reduction consisted of careful flat fielding and fringe removal to reduce all errors to less than 0.5%. More details are given in Bailer-Jones & Mundt [2].

3 Light Curve Analysis: Theory

Differential light curves were obtained for each target relative to a number of reference stars in the field. These reference stars were chosen to be bright and

Table 1. Properties of the ultra cool dwarf targets. Each reference makes use of a different I band and even definition of magnitude, so values are only intended to be indicative. In particular, the SDSS I filter is somewhat bluer than the Cousins I filter, thus yielding fainter magnitudes for L dwarfs. The spectral types in parentheses have been estimated from the $R - I$ colours in [5].

name	IAU name	I	SpT	Hα EW Å	LiI EW Å	ref
2M0030	2MASSW J0030438+313932	18.82	L2	4.4 ±0.2	< 1.0	[14]
2M0326	2MASSW J0326137+295015	19.17	L3.5	9.1±0.2	< 1.0	[14]
2M0345	2MASSW J0345432+254023	16.98	L0	≤ 0.3	< 0.5	[14]
2M0913	2MASSW J0913032+184150	19.07	L3	< 0.8	< 1.0	[14]
2M1145	2MASSW J1145572+231730	18.62	L1.5	4.2±0.2	< 0.4	[14]
2M1146	2MASSW J1146345+223053	17.62	L3	≤ 0.3	5.1±0.2	[14]
2M1334	2MASSW J1334062+194034	18.76	L1.5	4.2±0.2	< 1.5	[14]
2M1439	2MASSW J1439284+192915	16.02	L1	1.13±0.05	< 0.05	[25]
SDSS 0539	SDSSp J053951.99−005902.0	19.04	L5			[11]
SDSS 1203	SDSSp J120358.19+001550.3	18.88	L3			[11]
Calar 3		18.73	M9	6.5–10.2	1.8±0.4	[26]
Roque 11	RPL J034712+2428.5	18.75	M8	5.8±1.0		[30]
Roque 12		18.47	M7.5	19.7±0.3	≤ 1.5	[20]
Roque 16	RPL J034739+2436.4	17.79	M6	5.0±1.0		[30]
Teide 1	TPL J034718+2422.5	18.80	M8	3.5–8.6	1.0±0.2	[24]
S Ori 31	S Ori J053820.8−024613	17.31	(M6.5)			[5]
S Ori 33	S Ori J053657.9−023522	17.38	(M6.5)			[5]
S Ori 34	S Ori J053707.1−023246	17.46	(M6)	≤ 5.0		[5]
S Ori 44	S Ori J053807.0−024321	19.39	M6.5	60.0±1.0		[5]
S Ori 45	S Ori J053825.5−024836	19.59	M8.5			[5]
S Ori 46	S Ori J053651.7−023254	19.82	(M8.5)			[5]

isolated. Fluxes for all objects and reference stars were determined using aperture photometry with an aperture radius of 3.5 pixels (1.9″). The average flux of the reference stars forms a magnitude against which fluctuations in the target are monitored. The zero-mean differential light curve for each target, consisting of K points (or epochs), is denoted $m_d(1), m_d(2), \ldots, m_d(k), \ldots, m_d(K)$. The *total* photometric error at each point, $\delta m_d(k)$, has been carefully determined considering photon-noise statistics plus contributions from imperfect flat fielding and fringe correction.

3.1 χ^2 Test and Reference Star Rejection

A general test of variability is made using a χ^2 test, in which we evaluate the probability, p, that the deviations in the light curve are consistent with the photometric errors. The statistic is

$$\chi^2 = \sum_k^K \left(\frac{m_d(k)}{\delta m_d(k)} \right)^2 \qquad (1)$$

such that a large χ^2 indicates greater deviation compared to the errors, and thus a smaller p. An object is considered variable if $p < 0.01$ (a 99% confidence level).

This test is first used to iteratively remove variable reference stars, by calculating the light curve for each reference star relative to all the other reference stars. The most variable reference star is removed from the reference list, and the procedure repeated until only non-variable reference stars remain (i.e. those with $p > 0.01$). Due to the large number of reference stars used (typically 20–30), any residual low-level variability in any one object will be greatly reduced in the averaged reference level.

The reliability of the χ^2 test depends on an accurate determination of the magnitude errors in the target. This has been confirmed, as objects of similar brightness to the target have variations no larger than the photometric errors for the target, thus ensuring that the errors have not been overestimated.

3.2 Power Spectrum Analysis

Evidence for *periodic* variability was searched for using the *power spectrum* or *periodogram*. A dominant periodicity may be present at the rotation period due to rotational modulation of the light curve by surface inhomogeneities. The power spectrum of a continuous light curve, $g(t)$, is $|G(\nu)|^2$, where $G(\nu) = \mathrm{FT}[g]$ and FT denotes a Fourier transform. However, the target objects are only observed at the discrete time intervals given by the sampling function, $s(t)$. Thus the power spectrum of the measured *discrete* light curve, $d(t) = g(t)s(t)$, is $|D(\nu)|^2$, where $D(\nu) = \mathrm{FT}[d(t)] = G(\nu) \otimes W(\nu)$. $W(\nu) = \mathrm{FT}[s(t)]$ is the *spectral window function* and $\otimes$ is the convolution operator. Thus the measured power spectrum, $|D(\nu)|^2$, may show spurious features due to the way in which the true continuous light curve was sampled. Such features are not intrinsic to the source and may obscure features which are [10][27], particularly at the low SNRs considered here.

However, it is possible to estimate $G(\nu)$ (power and phases) from the raw or *dirty* power spectrum through an iterative deconvolution using the CLEAN algorithm, which was first introduced to reconstruct aperture synthesis data in radio astronomy [27]. The resulting *cleaned* spectrum generally consists of peaks at a number of distinct frequencies, plus a residual spectrum and noise. The CLEAN code used calculates the power and phases of the components in the window function and the cleaned power spectrum [18]. The noise in the power spectrum is calculated from the photometric errors and the time sampling, and is stated in the captions to the power spectra in the next section. Peaks which are not more than several times this noise level should not be considered significant. The uncertainty in a determined frequency in the power spectrum is set by $t_{\max}$, and is approximately $\tau^2/(2t_{\max})$ for a period τ. However, for very short periods the error is constant due to the finite integration time. The longest period to which the observations are sensitive is of order $t_{\max}$.

4 Results

4.1 General Results

The results of the application of the χ^2 test to the 21 targets are shown in Table 2 for the detections and Table 3 for the non-detections of variability. For those objects in which we did not detect variability, we have set upper limits on the amplitude. This was done by creating a set of synthetic light curves by multiplying each $m_d(k)$ by $1 + a$, for increasing (small) values of a. The amplitude limits were obtained from that synthetic light curve which gave $p = 0.01$ according to the χ^2 test. Note that a number of detections are close to the confidence limit of $p = 0.01$, so the division between Tables 2 and 3 is not a definitive statement of what is and what is not variable.

Table 2. Variability detections. $t_{\max}$ is the maximum time span of observations: the minimum span was between 10 and 20 minutes. Two measures of variability amplitude are given: the average of the absolute relative magnitudes, $\overline{|m_d|}$, and the root-mean-square (RMS) of the relative magnitudes, σ_m. $\overline{\delta m_d}$ is the average photometric error in the light curve. $1 - p$ is the probability that the variability is intrinsic to the target. "Obs. run" gives the date (YY-MM) of the observations.

target	SpT	$t_{\max}$ hours	$\overline{\|m_d\|}$ mags	σ_m mags	$\overline{\delta m_d}$ mags	p	No. of frames	No. of refs	Obs. run
2M0345	L0	53	0.012	0.017	0.011	4e-4	27	23	99-09
2M0913	L3	125	0.042	0.055	0.039	7e-4	36	14	99-01
2M1145	L1.5	124	0.026	0.031	0.022	1e-3	31	12	99-01
"	"	76	0.015	0.020	0.012	<1e-9	70	11	00-02
2M1146	L3	124	0.012	0.015	0.011	3e-3	29	7	99-01
2M1334	L1.5	126	0.017	0.020	0.011	<1e-9	51	12	00-02
SDSS 0539	L5	76	0.009	0.011	0.007	3e-5	31	24	00-02
SDSS 1203	L3	52	0.007	0.009	0.007	2e-3	51	13	00-02
Calar 3	M9	29	0.026	0.035	0.027	6e-4	42	21	99-01
S Ori 31	(M6.5)	50	0.010	0.012	0.007	4e-5	21	30	00-02
S Ori 33	(M6.5)	51	0.008	0.010	0.007	2e-3	21	43	00-02
S Ori 45	M8.5	50	0.051	0.072	0.032	5e-9	21	30	00-02

4.2 Comments on Individual Objects

Notes are now given on the light curves and power spectra of objects with statistically significant χ^2 detections. Brief comments are given at the end of the section on the non-detections. The implications of these results are discussed in section 5.

2M0345. The light curve shows no interesting features and there are no peaks in the cleaned power spectrum above four times the noise.

Surface Features, Rotation and Atmospheric Variability of Ultra Cool Dwarfs 97

Table 3. Variability non-detections. The columns are the same as in Table 2 except that here $\overline{|m_d|}$ and σ_m are the upper detection limits on the variability amplitudes. The minimum time between observations of a given target was between 3 minutes (for 2M1439) and 35 minutes (for Roque 12).

target	SpT	t_{max} hours	$\overline{|m_d|}$ mags	σ_m mags	$\overline{\delta m_d}$ mags	p	No. of frames	No. of refs	Obs. run
2M0030	L2	51	0.018	0.025	0.020	0.21	37	27	99-09
2M0326	L3.5	49	0.021	0.029	0.017	0.56	19	36	99-09
2M1439	L1	97	0.007	0.009	0.007	0.10	48	13	00-02
Roque 11	M8	100	0.028	0.043	0.027	0.46	47	23	99-01
Roque 12	M7.5	50	0.016	0.022	0.015	0.02	17	43	99-09
Roque 16	M6	29	0.010	0.014	0.010	0.35	16	34	99-09
Teide 1	M8	100	0.029	0.041	0.030	0.10	47	23	99-01
S Ori 34	(M6)	51	0.008	0.010	0.007	0.28	21	43	00-02
S Ori 44	M6.5	51	0.030	0.035	0.026	0.06	21	30	00-02
S Ori 46	(M8.5)	51	0.032	0.041	0.030	0.03	21	43	00-02

2M0913. This detection is due primarily to a significant drop in the flux around AJD 1187.5, going down to 0.13 magnitudes below the median for that night. There is no evidence for variability within the other three nights. There are no strong periodicities in the cleaned power spectrum, the strongest three being at 3.36, 0.76 and 0.64 (± 0.08) hours, each at around only five times the noise level.

2M1145. Evidence for variability in this L dwarf was presented in [1], and it was tentatively claimed to be periodic with a period of 7.1 hours (using the Lomb–Scargle periodogram), pending confirmation. These data (and that for all other targets in [1]) have been re-reduced, and the new reduction still shows significant evidence for variability. However, the 7.1 hour periodicity is no longer significant in the cleaned power spectrum. Peaks are present at $5.4\pm 0.1, 5.1\pm 0.1, 1.47\pm 0.08$ and 0.71 ± 0.08 hours, but are only marginally significant at around eight times the noise (Fig. 1). Note how difficult it would be to confidently identify these peaks in the dirty power spectrum.

The new reduction consists of an improved flat field, better fringe removal, more reference stars, and a few more points in the light curve. The light curves from the two reductions are consistent within their errors, but the power spectra differ, indicating that the 7.1 hour period was an artifact of higher noise and errors which crept into the first reduction.

2M1145 was re-observed at higher SNR and with more epochs in the 00-02 run. These data (Fig. 2) also show very strong evidence for variability, and the power spectrum shows significant peaks at the following periods (with power in units of the noise in parentheses): 11.2 ± 0.8 (31), 6.4 ± 0.3 (14), 2.78 ± 0.13 (7), 0.42 ± 0.13 (14) hours (Fig. 3). Note that the first period is four times the third, so these may not be independent. There are no common peaks in this power spectrum and the one from 99-01. This means that 2M1145 cannot have both stable surface features (over a one year timescale) and a rotation period of

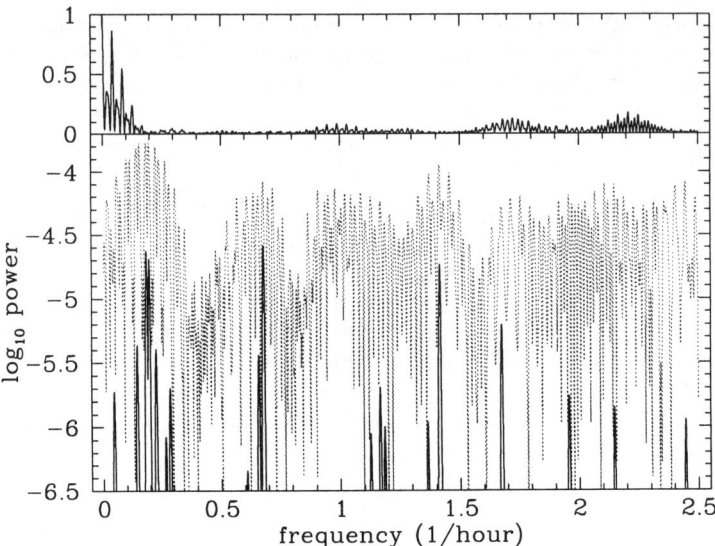

Fig. 1. Power spectrum for 2M1145 light curve from the 99-01 run. The bottom panel shows the dirty spectrum (dotted line) and the cleaned spectrum (solid line) in units of $\log_{10}(P)$. The noise level is at about $\log_{10}(P) = -5.6$. The top panel shows the shape of the spectral window function on a linear vertical scale, normalised to a peak value of 1.0.

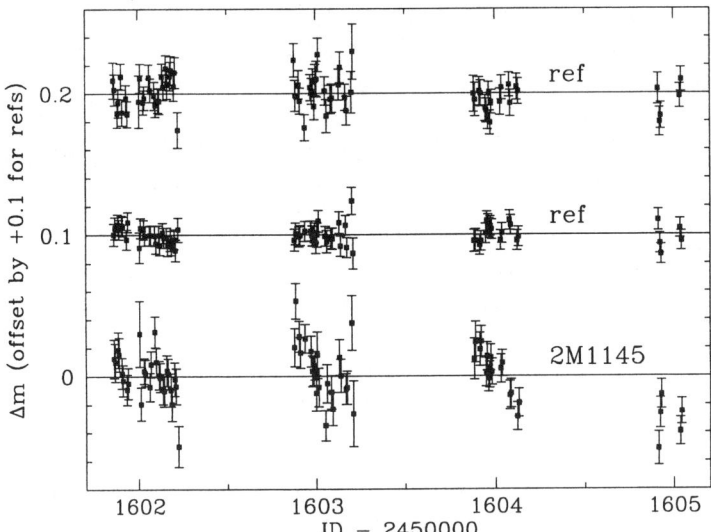

Fig. 2. Light curve for 2M1145 from the 00-02 run (bottom) plus a bright reference object (middle) and one of similar brightness to the target (top). The mean of these light curves each offset from the mean of the target star by the amount shown on the vertical axis. The mean for each light curve is shown as a solid line.

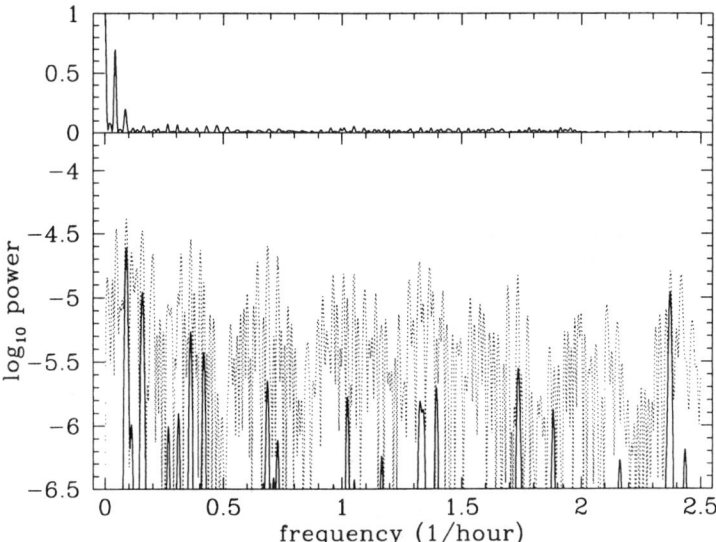

Fig. 3. Power spectrum for 2M1145 (from 00-02). The noise level is $\log_{10}(P) = -6.1$. See caption to Fig. 1.

between 1 and 70 hours; if it did we would have detected it in both runs (see section 5.1).

2M1146. The power spectrum shows peaks at the following periods (with power in units of noise): 5.1 ± 0.1 (15), 3.00 ± 0.08 (6), 1.00 ± 0.08 (5), and 0.64 ± 0.08 (9) hours. The second and third are in the ratio 3:1, so are probably not independent. The one at three hours is more convincingly based on the phase coverage in the phased light curve. This is one of only two L dwarfs in the sample which has a measured $v \sin i$, which, at 32.5 ± 2.5 km/s [4], implies a rotation period of 3.7 ± 0.3 hours, or less, due to the unknown inclination of the rotation axis (assuming a radius $0.1 R_\odot$ [8]). 2M1146 appears to be an L dwarf–L dwarf binary [16] with a separation $0.3''$, as well as having a background early-type star only 1" away [14]. The light curve is a composite of variations in all three objects.

2M1334. This is significantly variable, and the light curve shows clear fluctuations within a number of nights (Fig. 4). The largest peak in the power spectrum (Fig. 5) is at 2.68 ± 0.13 hours at 12 times the noise.

Calar 3. The light curve shows no conspicuous features. The two most significant peaks in the power spectrum (at 14.0 and 8.5 hours) are less than five times the noise level, so are barely significant.

SDSS 0539. The seeing was worse than average for many of the frames in this field, so a larger photometry aperture of radius 5.0 pixels was used. (This in-

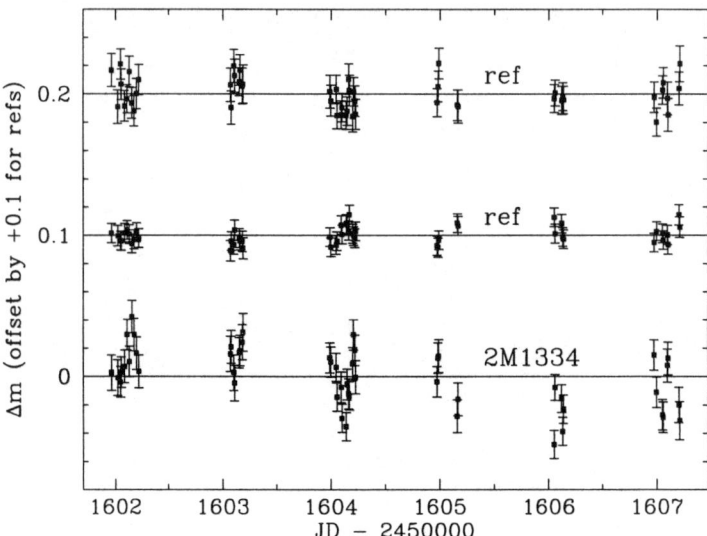

Fig. 4. Light curve for 2M1334 (bottom) plus a bright reference object (middle) and one of similar brightness to the target (top). See caption to Fig. 2.

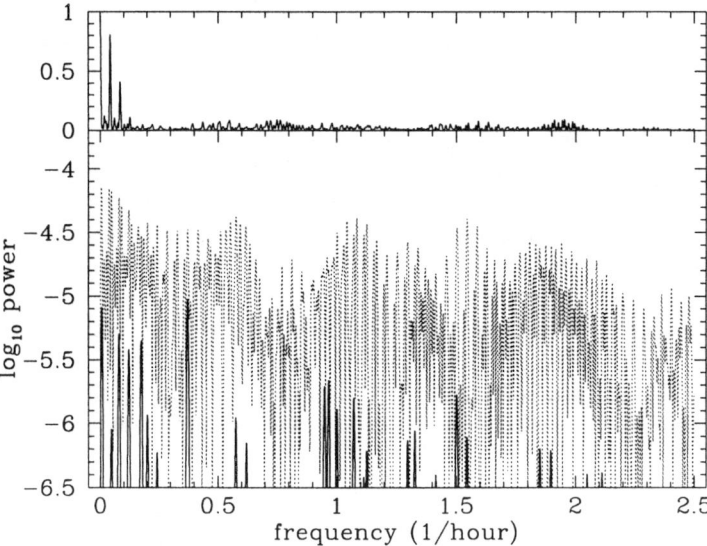

Fig. 5. Power spectrum for 2M1334. The noise level is $\log_{10}(P) = -6.2$. See caption to Fig. 1.

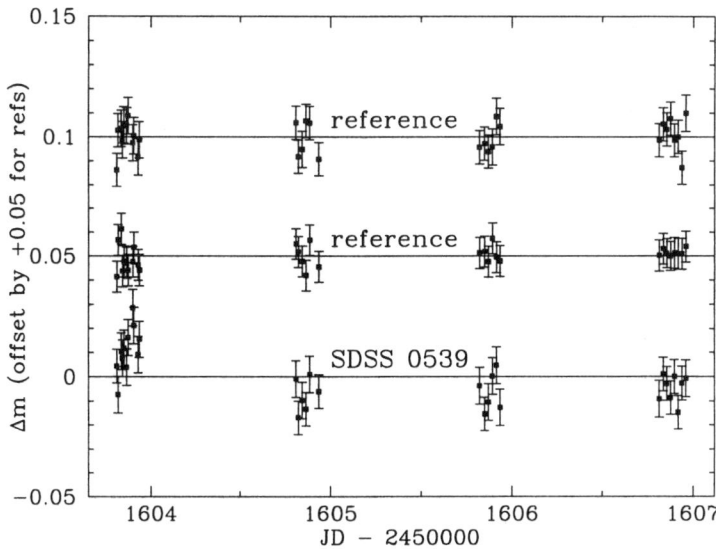

Fig. 6. Light curve for SDSS 0539 (bottom) plus a bright reference object (middle) and one of similar brightness to the target (top). See caption to Fig. 2.

creases the noise and hence lowers the sensitivity.) The significant χ^2 is partly due to the brighter points around AJD 1604. Otherwise the light curve shows no obvious patterns (see Fig. 6). The power spectrum shows a significant (20 times noise) peak at 13.3 ± 1.2 hours (Fig. 7). The light curve phased to this period is shown in Fig. 8.

SDSS 1203. This variability is primarily due to a drop in brightness of about 0.02 magnitudes in four consecutive measurements around AJD=1606.1 (Fig. 9) lasting between one and two hours. This could be due to a short-lived surface feature, or possibly an eclipse by a physically associated companion.

S Ori 31. The power spectrum shows significant peaks at 7.5 ± 0.6 and 1.75 ± 0.13 hours at 18 and 9 times the noise level respectively. The former period dominates and may be the rotation period.

S Ori 33. The light curve shows a rise just before AJD 1606, and the power spectrum has peaks of 6 to 7 times the noise at 8.6 ± 0.7 and 6.5 ± 0.4 hours. The former has good phase coverage, so may be the rotation period, although it is not a strong peak.

S Ori 45. The light curve shows three points much lower than the average around AJD 1604.9, spanning a range of almost 0.25 magnitudes. There is, however, a bright ($\Delta m = 1.7$) nearby (5″) star which may well interfere with this variability determination. If these points are excluded there is no evidence for variability

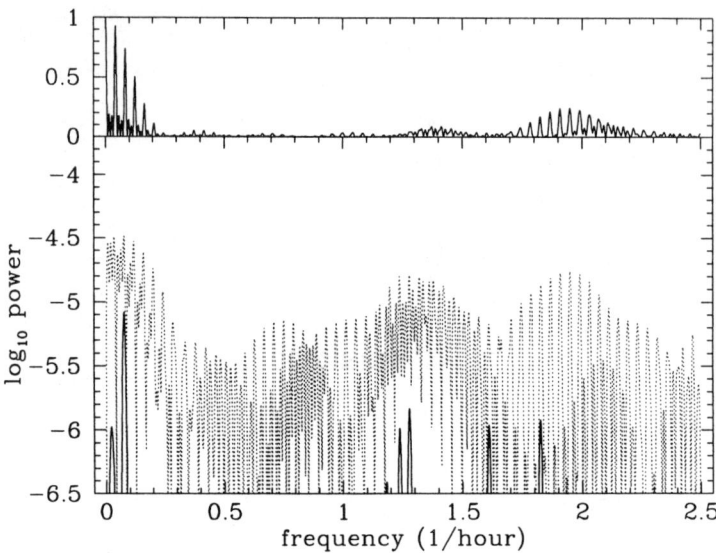

Fig. 7. Power spectrum for SDSS 0539. The noise level is $\log_{10}(P) = -6.4$. See caption to Fig. 1.

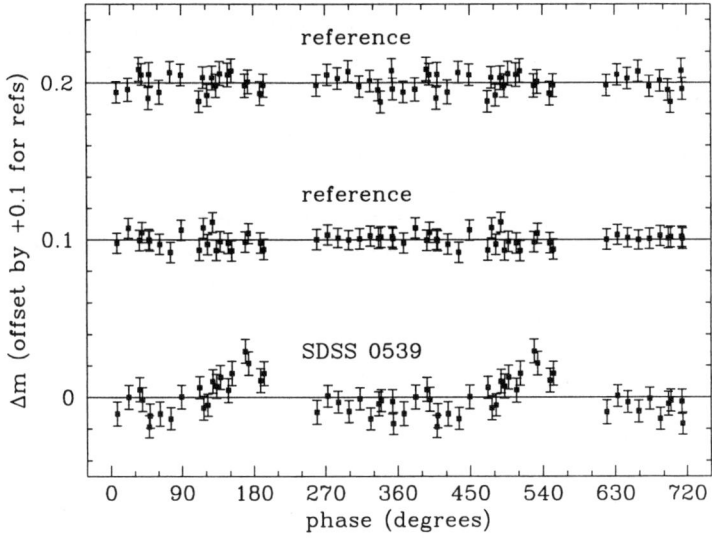

Fig. 8. Light curve (bottom) for SDSS 0539 phased to a period of 13.3 hours. The cycle is shown twice (°–360° and 360°–720°). Also shown are two reference stars from Fig. 6 phased in the same way.

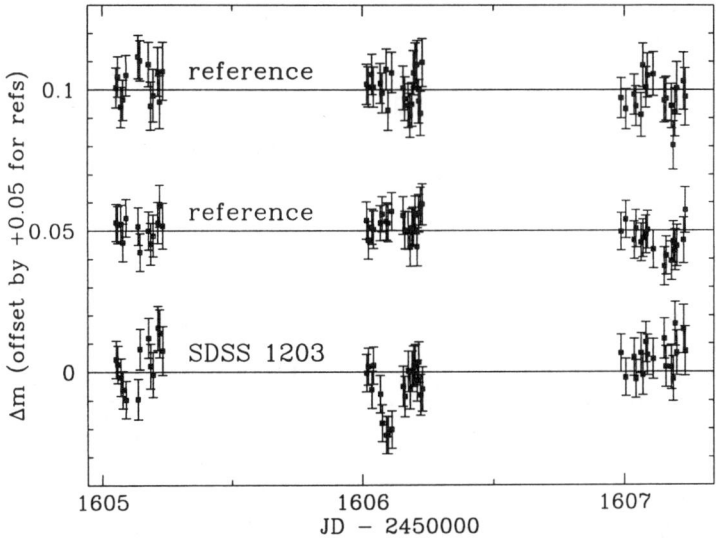

Fig. 9. Light curve for SDSS 1203 (bottom) plus a bright reference object (middle) and one of similar brightness to the target (top). See caption to Fig. 2.

($p = 0.18$). The most significant peak in the power spectrum is at 0.50 ± 0.13 hours (at 20 times the noise), which would be extremely fast if it is the rotation period.

Non-detections. 2M1439 has been measured to have a $v \sin i$ of 10 ± 2.5 km/s [4], implying a period of less than 12.1 hours for a $0.1 R_\odot$ radius. S Ori 46 has a bright nearby star, which may affect the attempt to determine variability in this object. Roque 11 and Teide 1 have also been observed for variability in the I band by Terndrup et al. [28]. They also did not find evidence for variability, with measured values of σ_m of 0.041 and 0.045 magnitudes respectively.

4.3 Summary of the Results

11 of the 21 objects show evidence for variability at the 99% confidence level ($p = 0.01$). Of these, four (2M1145, 2M1334, SDSS 0539, S Ori 31) show strong evidence for variability ($p <$ 1e-4). S Ori 45 is formally a fifth object with strong evidence for variability, but the presence of a bright close star makes us hesitant to draw this conclusion. In four cases (2M1146, 2M1334, SDSS 0539, S Ori 31) we have detected dominant significant periods in the range 3–13 hours, which may be rotation periods in all but the first case. S Ori 45 also has a dominant peak, but at 0.5 hours this would be very rapid if it is a rotation. The remaining objects do not show dominant periods, although the two earliest-type variables (S Ori 31 and S Ori 33) show near-sinusoidal light curves at detected periods. The light curve of one object, SDSS 1203, is essentially featureless except for a

dip which may be due to an eclipse by a companion, although there is no direct evidence for this.

All of the objects which show variability have RMS amplitudes (σ_m in Table 2) between 0.01 and 0.055 magnitudes (ignoring S Ori 45), but most lie in the range 0.01 to 0.03 magnitudes and vary on timescales of a few hours. More detailed results are provided in Bailer-Jones & Mundt [2].

5 Discussion

5.1 Simulations of the Light Curves of Rotating Spotted Stars

The power spectrum is a representation of the light curve in the frequency domain. Specifically it gives the contributions to the variance in the light curve of sinusoids as a function of their frequencies. However, a significant peak in the power spectrum does necessarily correspond to a *long-term* (and hence meaningful) periodicity. After all, *any* light curve – including a random one – can be described in terms of its power spectrum, as all features in the time domain must appear in the frequency domain somehow.

In particular, the "ideal" case of a pure sinusoidal light curve is only produced by a rotating star if one hemisphere is uniformly darker than the other and the star is observed along its equatorial plane. In contrast, a star with a single small surface feature ("spot") would show a sinusoidal light curve only when the spot is on the observable hemisphere; for up to half of the rotation (depending on the inclination of the rotation axis) the light curve would be constant. A star with several spots will show a more complex behaviour, due to the variable number of spots observable (and hence modulating the light curve) at any one time. All of these variations will be explained by apparent "periods" in the power spectrum, some of which may even be significant relative to the noise.

We have carried out numerous simulations to understand the appearance of the light curve and its power spectrum due to such spots. Fig. 10 shows the light curve due to a single small dark spot on a star. If we rotate this star with a period of five hours and observe it with the same noise level and time sampling as one of the targets (2M1334), we obtain the power spectrum and phased light curve in Fig. 11 and Fig. 10C respectively. The phased light curve is not sinusoidal, yet the power spectrum detects the rotation period.

Another simulation is shown in Fig. 12, which is due to a star with eight spots rotating with a period of ten hours. Here the contrast of the individual spots is much smaller, only -0.008 to $+0.014$ magnitudes. The sampling and noise from 2M1145 (00-02 run) is used and results in a significant variability detection according to the χ^2 test, but with $p=0.005$ is close to the variable/non-variable cut-off. Despite this low SNR (and no detectable sinusoidal variation in Fig. 12C), the rotation period still clearly stands out in the cleaned power spectrum (Fig. 13).

We see that the light curve phased to the detected rotation period does not necessarily show sinusoidal (or even near-sinusoidal) variation. More extensive

Surface Features, Rotation and Atmospheric Variability of Ultra Cool Dwarfs 105

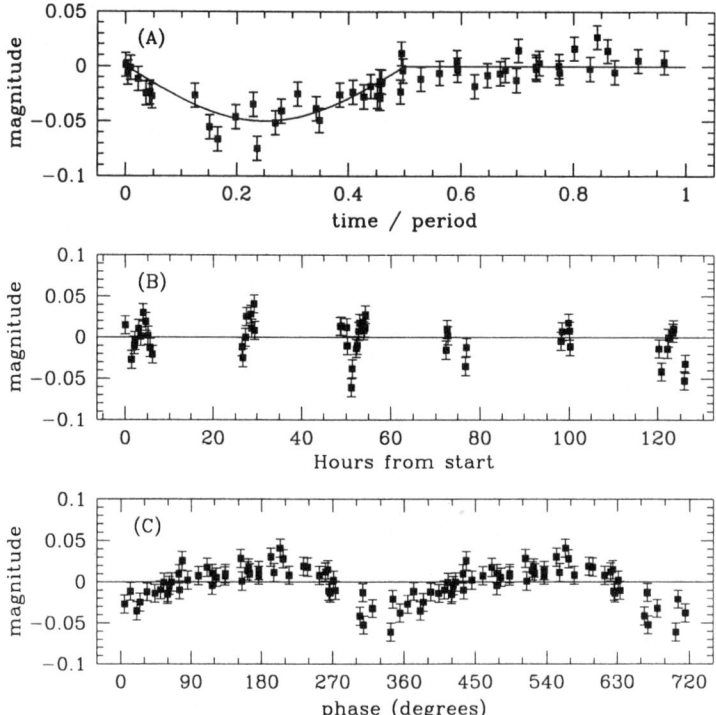

Fig. 10. (A) The solid line shows the true (noiseless) light curve of a rotating star viewed equatorially with a single dark spot of 0.05 magnitudes contrast. The star is rotated with a period of five hours and observed as 2M1334 was (i.e. with the same time sampling and with additive Gaussian noise of standard deviation 0.011 mags), giving the observed light curve in (B). (These points are also plotted in (A) wrapped to the rotation period.) Thus is significantly variable according to the χ^2 test ($p < 1$e-9). The cleaned power spectrum (Fig. 11) detects a period at 5.01 ± 0.10 hours: the light curve phased to this detected period *and phase* is shown in (C) (cycle shown twice).

simulations with different numbers, amplitudes and phases of spots have been carried out. These indicate that even if the contrast of the spots (and hence the SNR) is very low, then provided the light curve shows significant variation according to the χ^2 test, the rotation period is seen at more than ten times the noise in the power spectrum. Thus the absence of significant periods in the power spectrum for a variable light curve indicates *non-periodic* variations over the timescale of the observations.

5.2 Evidence for the Evolution of Surface Features

This non-periodic variability appears to imply one of three things:

1. the rotation period is shorter than the time span of observations
2. the rotation period is longer than the time span of observations

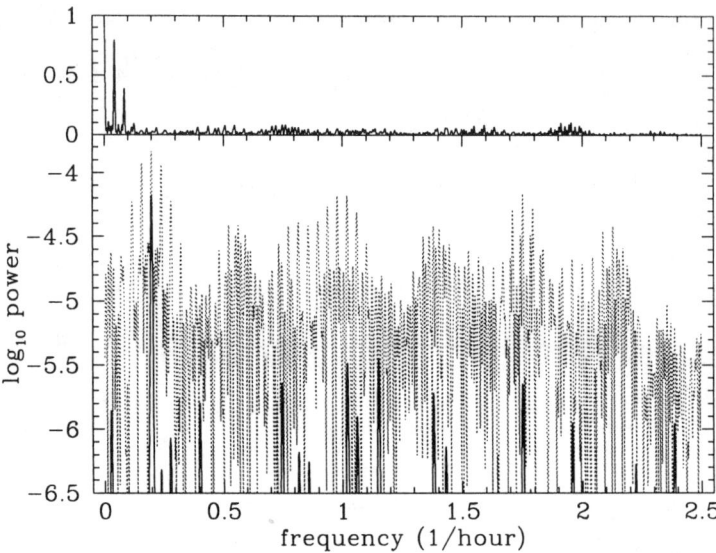

Fig. 11. Power spectrum for the simulated light curve shown in Fig. 10B. The noise level is $\log_{10}(P) = -6.2$. The same CLEAN parameters were used here as for the real data of section 4. See caption to Fig. 1.

3. the surface features which are presumed to be modulating the light curve are not stable over the time span of observations.

The first of these implies a rotation period of less than 0.4 hours, which corresponds to an equatorial rotation speed of at least 240 km/s. Based on t_{max} in Tables 2 and 3, the second possibility requires *maximum* $v \sin i$ values (i.e. when viewed equatorially) of between 1 and 4 km/s, assuming a radius of $0.1 R_\odot$. These would be inconsistent with the measurements of Basri et al. [4], who report $v \sin i$ values between 10 and 60 km/s for 16 out of 17 late M and L field dwarfs. (According to models, even the youngest, warmest objects in Table 1 – those in σ Orionis – can have radii no larger than $0.2 R_\odot$ [8], so $v \sin i$ could not be above 8 km/s for periods of order t_{max}.) The results of Basri et al. therefore imply typical rotation periods of 1 to 10 hours, and the simulations have shown that such periods would have been detected in the light curves of the present sample, *if* these objects had stable modulating surface features. Yet some significantly variable objects show no significant periodicities. The logical explanation in these cases (especially 2M0345, 2M0913, 2M1145 and Calar 3) is, therefore, that these objects have surface features which evolve over the period of the observations, thus removing the rotational modulation from the light curve. For 2M1145 we possibly have more direct evidence of this, as the two light curves from one year apart show no common periods.

Surface Features, Rotation and Atmospheric Variability of Ultra Cool Dwarfs   107

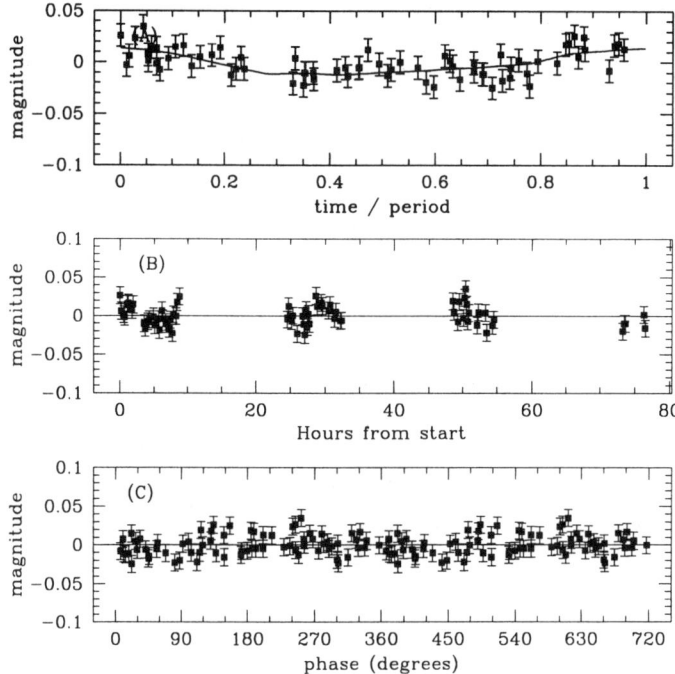

Fig. 12. Same as Fig. 10 except now for eight dark and bright spots with random phases. This gives a significant detection, although not overwhelming ($p = 0.005$), yet the cleaned power spectrum (Fig. 13) still detects the rotation period of 10 hours.

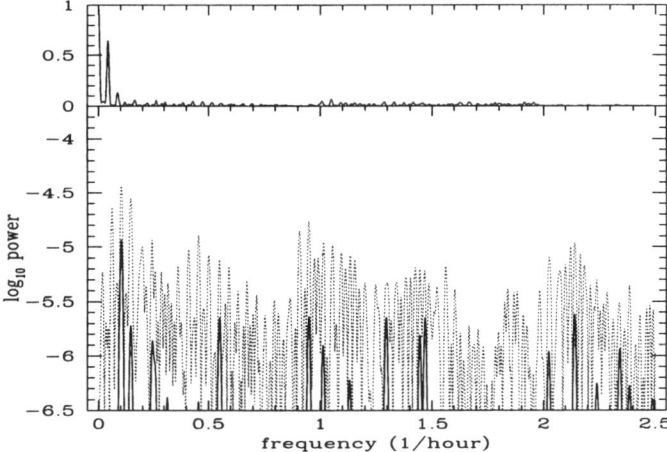

Fig. 13. Power spectrum for the simulated light curve shown in Fig. 12B. The noise level is $\log_{10}(P) = -6.1$. See caption to Fig. 1.

5.3 Physical Nature of the Surface Features

Magnetically-induced star spots are common in solar-type stars, the magnetic field being produced by the $\alpha\Omega$ dynamo. This appears not to operate in low mass stars and brown dwarfs, yet a turbulent dynamo may [8]. However, recent observations imply that chromospheric activity – and, perhaps, the contrast of magnetic spots – decreases rapidly between spectral types M7 and L1 [3][12]. In comparison, Fig. 14 shows the amplitude of variability (or upper limit thereon) as a function of spectral type for the sample in this paper. Whereas 7/10 of objects later than M9 show variability, only 2/9 earlier than this do. (The average detection limits/amplitudes are almost identical in the two regions, so this is not an artifact.) If the variability were due to magnetic spots, then in the light of the activity decline we would expect variability to be *less* common among later-type objects, not *more* common as seen here. Although this could also be an age effect (all of the targets of type M9 and earlier are cluster members with ages less than 120 Myr) it hints towards a non-magnetic origin of the surface features.

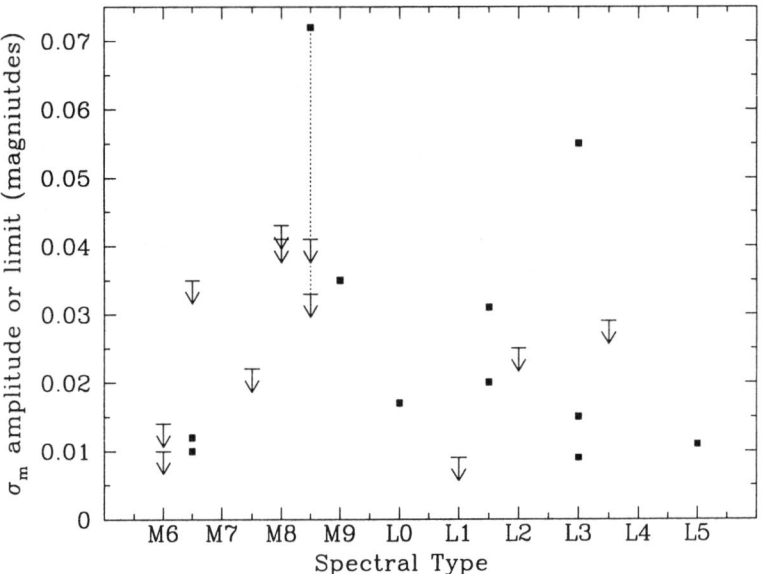

Fig. 14. Relationship between variability amplitudes (squares) or upper limits to variability (arrows) and spectral type. S Ori 45 (M8.5) is plotted as both an amplitude and a limit (connected with a dotted line) depending on whether the first night of data is included or not. The plot using $\overline{|m_d|}$ rather than σ_m as the amplitude measure is very similar.

Another candidate for producing variability is photospheric dust clouds. Modelling of optical and infrared spectra show that solid particles form in late M and L dwarfs [7][9][13][19]. While we may expect this dust to gravitationally settle below the photosphere, certain processes (such as tubulence) may prevent this, and models which include dust opacity (as well as removing dust constituents from the equation of state) give better fits to the near infrared spectra of ultra cool dwarfs [9]. This dust could form into large-scale opaque clouds. Their evolution over a few rotation periods could account for the variability reported in this paper, possibly driven by rapid rotation and turbulence. The clouds would have to be relatively large, as many small clouds evolving independently would have an insignificant net effect on the light curve. As more dust can form in cooler objects, we may expect more variability in later-type objects, as seen in Fig. 14. However, given the small amount of data on any one object, it is premature to attempt to determine the characteristics or dynamics of the variability patterns.

Other causes of the variability can be considered, including flaring, accretion (for the youngest objects) or infall in an interacting binary. However, these all rely on transient phenomena, so no one explanation for all objects is that satisfactory.

6 Summary

Light curves for 21 late M and L dwarfs were obtained to probe variability on timescales between a fraction of an hour and over 100 hours. 11 objects showed evidence for variability at the 99% confidence level according to a χ^2 test, with amplitudes between 0.009 and 0.055 magnitudes (RMS). The ten non-detections have upper limits on their RMS amplitudes of between 0.009 and 0.043 magnitudes.

Power spectral analysis showed that a few objects (namely 2M1146, 2M1334, SDSS 0539 and S Ori 31) had significant, dominant periods between 3 and 13 hours. For 2M1334, SDSS 0529 and S Ori 31 these may be the rotation periods. The remaining seven significantly variable light curves did not show dominant periods, and in at least three cases (2M0345, 2M0913, Calar 3) there are not even any significant periods. Simulations showed that any plausible period would have been detected for these objects. It was concluded that this non-periodic behaviour is probably due to the evolution of surface features (assumed to produce the variability) on timescales of a few to a few tens of hours. These variabilities blur the rotation period thus inhibiting its detection. This is supported by observations of 2M1145 one year apart, in which the two light curves have no common periodicities.

It was postulated that this variability (at least for the later-type objects) is due to photospheric dust clouds, the evolution of which could be driven by turbulence and rapid rotation. In support of this is the greater propensity for variability in objects later than M9, while magnetic activity (which could otherwise support the presence of magnetically-induced spots) declines greatly beyond M7.

Acknowledgements

The author would like to thank Harry Lehto for use of his CLEAN code and advice on its use. The data in this paper were obtained with the 2.2m telescope at the German–Spanish Astronomical Center at Calar Alto in Spain.

References

1. C.A.L. Bailer-Jones, R. Mundt: A&A 348, 800 (1999)
2. C.A.L. Bailer-Jones, R. Mundt: A&A submitted (2000)
3. G. Basri, Cool Stars 11 (ASP Conf. Ser.), in press (2000)
4. G. Basri, S. Mohanty, F. Allard, et al.: ApJ 538, 363 (2000)
5. V.J.S. Béjar, M.R. Zapatero Osorio, R. Rebolo: ApJ 521, 671 (1999)
6. A.J. Burgasser, J.D. Kirkpatrick, M.E. Brown, et al.: AJ 120, 1100 (2000)
7. A. Burrows, C.M. Sharp: ApJ 512, 843 (1999)
8. G. Chabrier, I. Baraffe: ARAA, in press (2000)
9. G. Chabrier, I. Baraffe, F. Allard, P. Hauschildt: ApJ 542, 464 (2000)
10. T.J. Deeming: Astrophysics & Space Science 36, 137 (1975)
11. X. Fan, G.R. Knapp, M.R. Strauss, et al.: AJ 119, 928 (2000)
12. J.E. Gizis, D.G. Monet, I.N. Reid, J.D. Kirkpatrick, J. Liebert, R.J. Williams: AJ 120, 1085 (2000)
13. H.R.A. Jones, T. Tsuji: ApJ 480, L39 (1997)
14. J.D. Kirkpatrick, I.N. Reid, J. Liebert, et al.: ApJ 519, 802 (1999)
15. J.D. Kirkpatrick, I.N. Reid, J. Liebert, et al., AJ 120, 447 (2000)
16. D.W. Koerner, J.D. Kirkpatrick, M.W. McElwain, N.R. Bonaventura: ApJ 526, L25 (1999)
17. S.L. Leggett, T.R. Geballe, X. Fan, et al.: ApJ 536, L45 (2000)
18. H.J. Lehto: private communication (2000)
19. K. Lodders: ApJ 519, 793 (1999)
20. E.L. Martín, G. Basri, M.R. Zapatero-Osorio, R. Rebolo, R.J. García López: ApJ 507, 41 (1998)
21. E.L. Martín, X. Delfosse, G. Basri, B. Goldman, T. Forveille, M.R. Zapatero-Osorio: ApJ 118, 2466 (1999)
22. T. Nakajima, T. Tsuji, T. Maihara, et al.: PASJ 52, 87 (2000)
23. K.S. Noll, T.R. Geballe, S.K. Leggett, M.S. Marley: ApJ 541, L75 (2000)
24. R. Rebolo, M.R. Zapatero Osorio, E.L. Martín: Nature 377, 129 (1995)
25. I.N. Reid, J.D. Kirkpatrick, J.E. Gizis, et al.: AJ 119, 369 (2000)
26. R. Rebolo, E.L. Martín, G. Basri, G.W. Marcy, M.R. Zapatero-Osorio: ApJ 469, L53 (1996)
27. D.H. Roberts, J. Léhar, J.W. Dreher: AJ 93, 968 (1987)
28. D.M. Terndrup, A. Krishnamurthi, M.H. Pinsonneault, J.R. Stauffer: ApJ 118, 1814 (1999)
29. C.G. Tinney, A.J. Tolley: MNRAS 304, 119 (1999)
30. M.R. Zapatero Osorio, R. Rebolo, E.L. Martín, et al.: A&AS 134, 537 (1999)

Low-Mass Stellar and Brown Dwarf Binary Systems

I.N. Reid[1], D.W. Koerner[1], J.E. Gizis[2], and J.D. Kirkpatrick[2]

[1] University of Pennsylvania, Department of Physics & Astronomy,
209 South 33rd Street, Philadelphia, PA 19104, USA
[2] IPAC, Caltech, Pasadena, CA 91125, USA

1 Introduction

Binary systems have long been recognised as providing one of the most powerful methods of probing the physical properties of stars drawn from the full range of the HR diagram. Most notably, the overwhelming majority of measurements of mass, the fundamental stellar parameter, rest on observations of astrometric or spectroscopic binaries. For main sequence stars, those individual measurements can be combined to define a mass-luminosity relation which, applied to a suitably-constructed sample, permits derivation of $\Psi(M)$, the mass function. The latter, in its turn, provides a valuable constraint on star formation theory by defining the end product of an assortment of processes, including collapse, fragmentation, coagulation and accretion, which combine to transform gas into stars. In similar fashion, the frequency of binaries as a function of mass and the distribution of mass ratios and separations offer important clues to the underlying mechanism(s) which conspire reduce individual molecular cloud cores to to gaseous spheres in hydrostatic equilibrium.

Almost all of these considerations apply with equal force to substellar mass objects - brown dwarfs, starlike objects which fail to generate sufficient energy on collapse to drive the central temperature above the critical value for sustained hydrogen fusion. The exception lies in the calibration of $\Psi(M)$: since brown dwarfs fail to sustain hydrostatic equilibrium (they are degenerate), the mass-luminosity relation is not single valued. Brown dwarfs of all masses follow almost identical tracks, but at a rate which decreases with decreasing mass. Individual mass determinations, however, are still of considerable importance in matching against the predictions of theoretical models, and brown dwarf statistics are as important as stellar studies in characterising star formation.

The identification of substellar binaries is an obvious prerequisite to either individual or statistical analysis. This short paper outlines some results in this area, concentrating on recent discoveries based on high-resolution imaging with the Hubble Space Telescope. Those observations, combined with data from other sources spanning a larger mass range, suggest both a preference for equal-mass systems amongst very low-mass (VLM) dwarfs and an intriguing correlation between the maximum separation of binary systems and the total system mass.

2 L Dwarf/L Dwarf Binary Systems

Any analysis of binary statistics must carry as subheadings both the range of separations and luminosity differences appropriate to the contributing technique or techniques and the range of primary mass covered. Too often one encounters superficial studies which, willy-nilly, apply results from Duquennoy & Mayors [3] G dwarf survey, regardless of their relevance. In particular, Fischer & Marcy [4] quickly established that M dwarfs have a significantly lower binary fraction than the higher-mass solar-type stars, 35%vs. 60%[1]. That value is clearly more relevant as a reference for the current summary, centred on primary masses $M < 0.1 M_\odot$.

All currently-known, confirmed L dwarf/L dwarf binary systems are spatially resolved, identified through high-resolution imaging[2] Ground-based near-infrared imaging was responsible for identifying Denis1228 [10], Denis0205 and 2M1146+2230 [8] as binaries, each with components of near-equal luminosity. Optical spectroscopy shows that both Denis1228 and 2M1146 possess a strong Li I 6708Å absorption line, indicative of the presence of a significant abundance of primordial lithium. Since dwarfs are known to become fully convective at temperatures below ~ 3300K, the continued survival of this element implies that the central temperature barely reached $T_C \sim 2 \times 10^6$K at any point in the evolutionary history. That upper limit falls below the critical value for sustained H-burning [14], identifying Denis1228 and 2M1146 as *bona fide* brown dwarfs. As already noted in the introduction, the rate of evolution of a brown dwarf depends on its mass; thus, if we make the not unreasonable assumption that binary components are coeval, near-equal luminosity implies near-equal mass.

Koerner et al. [8] (K99) provide the first statistical analysis of binary frequency in L dwarfs, compiling Keck near-infrared observations of ten L dwarfs. Three (named above) are binary, including two of the three L dwarfs discovered in the initial DENIS brown dwarf mini-survey [2]. Taken at face this, this relatively high discovery rate at separations exceeding ~ 5 AU suggests a *higher* binary frequency than amongst M dwarfs, as well as a possible preference for equal-mass systems, the latter echoing results for local M dwarfs [16].

However, there is a caveat to this result. Koerner et al.'s analysis centres on some of the brightest known L dwarfs, and is therefore a magnitude-limited, rather than volume-limited, sample. Öpik [13] originally pointed out that, since binaries with equal-luminosity components are twice as bright as single stars of the same spectral type, they have the same apparent magnitude at distances larger by a factor of $\sqrt{2}$. In a magnitude-limited sample, this translates to a search volume larger by a factor of $2\sqrt{2}$, so a higher detection rate of such systems is not necessarily indicative of an increased frequency.

---

[1] Since 80% of all stars are M dwarf, this is a more accurate representation of *overall* stellar binarity than the oft-quoted 50-60%.

[2] Reid et al. [17] originally proposed 2M0345+2540 as a double-lined spectroscopic binary, based on HIRES observations, but subsequent analysis casts doubt on that interpretation. High resolution NIRSPEC observations should resolve this issue.

We have addressed this issue through HST Planetary Camera observations of a much more extensive sample of L dwarfs. Our cycle 8 observations are drawn from a target list of 37 dwarfs, selected from the 2MASS samples identified by Kirkpatrick et al. [6], [7]. In addition, we are currently undertaking a Cycle 9 Snapshot proposal in conjunction with members of the SDSS consortium, targetting 120 ultracool dwarfs drawn from both 2MASS and SDSS. Neither of these samples is strictly volume limited, but both span a much larger range of apparent and absolute magnitude than the initial K99 sample. As a result, Öpik's equal-luminosity binary bias should be much less of a factor.

Fig. 1. The brown dwarf binary, 2M0850+1037. The bright star in the upper left is an unrelated M dwarf.

HST Snapshot observations were obtained for twenty of our Cycle 8 targets. A full discussion of the results is given by Reid et al. [18]; in brief, only four are resolved as binary, all with separations $\Delta < 10$ AU. Three of the four have components with near-equal luminosity; in the fourth, 2M0850+1037, the secondary is fainter than the primary by 1.7 magnitudes in the I-band, and is ~ 0.5 magnitudes fainter in absolute terms than the lowest luminosity known L8 dwarf (Fig. 1). Indeed, 2M0850B is only ~ 1.5 magnitudes brighter at I than the T

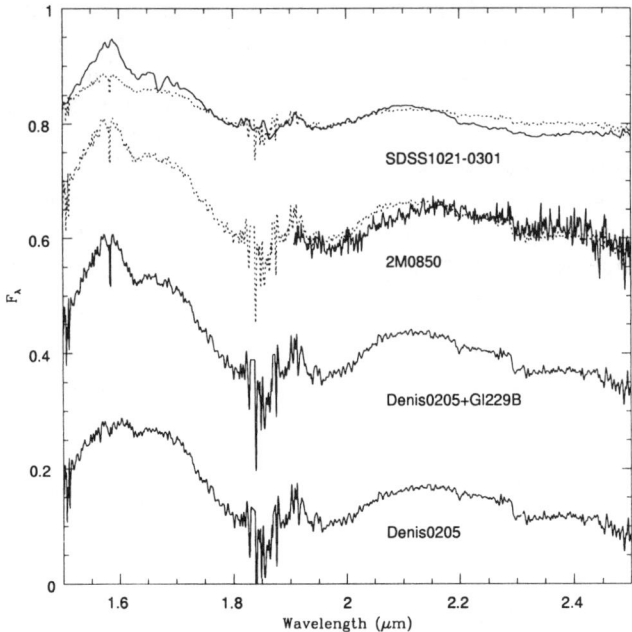

Fig. 2. A comparison between the observed K-band spectrum of 2M0850+1037, an early T dwarf (identified from SDSS data) and an hypothetical L dwarf/T dwarf binary.

dwarf archetype, Gl 229B. Since the primary has significant lithium absorption, we know that both components have masses below $\sim 0.06 M_\odot$, and a comparison with models indicates a mass ration of $q \approx 0.8$, with the primary likely to have $M \sim 0.055 \pm 0.005 M_\odot$.

The low luminosity of the secondary in this system raises the possibility that the system consists of an L dwarf/T dwarf pair. Infrared spectroscopy can be used to test this hypothesis. Based on the relative luminosities, we can estimate that $\frac{F_1}{F_2} \sim 3$ in the J-band; scaling the Gl 229B flux distribution appropriately, this implies $\frac{F_1}{F_2} \sim 10$ at K. However, almost all of the K-band flux is emitted in the shorter wavelength half of the passband, since CH_4 absorption removes the longer-wavelength radiation. As a result, the energy distribution of a composite L dwarf/T dwarf binary with this luminosity difference is affected significantly, as illustrated in Fig. 2. It seems clear that 2M0850 does not exhibit the requisite attributes, although H-band spectroscopy would provide a definitive answer.

The overall energy distribution of the composite L dwarf/T dwarf binary is reminiscent of the early T dwarfs identified by SDSS [9]. Given the JHK colours of those dwarfs, it is unlikely that any are unresolved binaries; however, one should note that Öpik's criterion would lead one to expect a number of such systems amongst the brightest objects with that spectral energy distribution.

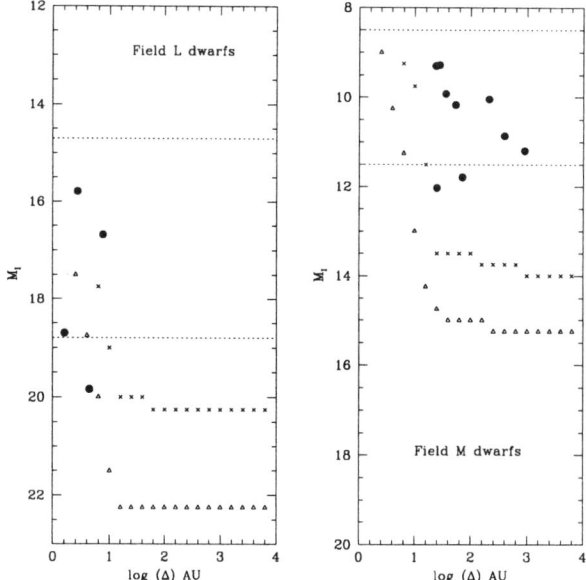

Fig. 3. Magnitude difference as a function of linear separation for binaries detected in our HST L dwarf survey and the RG97 field M dwarf survey. The triangles outline the effective detection limits as a function of linear separation. Note the smaller separations of the L dwarf/L dwarf binaries

3 Statistics of VLM Binary Systems

We noted previously that the three binary L dwarfs identified by K99 all have linear separations of less than 10 AU. The same holds for the four binaries identified amongst our HST PC sample (there is one system, 2M1146, in common between the two datasets). This result does not square well with the expected results that L dwarf binary systems should follow the same distribution of separations as M dwarfs. Figure 3 illustrates the problem: Reid & Gizis [16] (RG97) obtained HST PC images of 55 field M dwarfs. Despite the average distance of 55 parsecs, as opposed to the 25-30 parsecs for the L dwarfs, the overall fraction of *detected* binaries is almost identical: $17\pm5\%$ for the M dwarfs versus $20\pm10\%$ for the L dwarfs. However, the M dwarf data sample significantly larger separations: none of the L dwarf binaries would be detectable amongst the field stars; conversely, there is no evidence for any L dwarf systems at separations matching those of field M dwarfs. Combined with other observational data, notably a survey of Pleiades VLM dwarfs [11] and Koerner et al.'s NIRC data for 60 VLM dwarfs, these data point to a significant deficit of VLM/VLM dwarf binaries at separations exceeding 10 AU.

That is not to say that there no brown dwarfs as wide companions in binary systems; two obvious examples are GD 165B, the first known L dwarf [1], and Gl

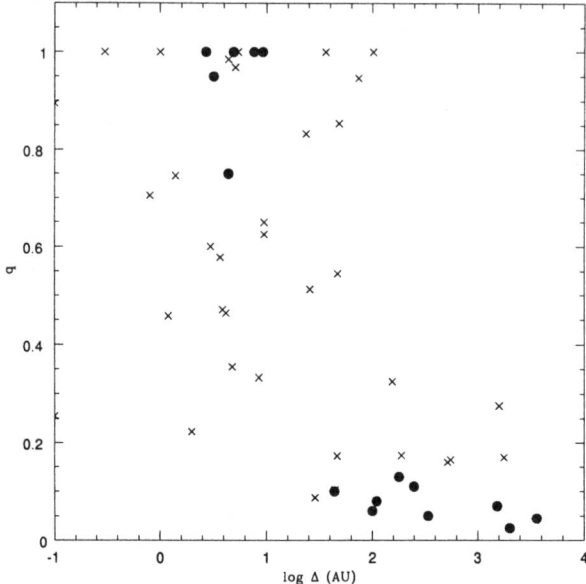

Fig. 4. The mass ratio, q, as a function of separation for low-mass binaries. L dwarf & brown dwarf binaries are plotted as solid points; nearby M dwarfs as crosses.

229B, the first unequivocal brown dwarf [12]. However, in each of those cases, as with G196-3B, Gl 584C and Gl 570D, the primary is substantially more massive than the VLM secondary. Other examples are given by Kirkpatrick et al. (this volume). Indeed, calculations suggest that brown dwarfs are found as frequently as M dwarfs as companions in wide binaries ($\Delta > 100$AU) (Gizis et al., in prep.). This leads to an apparently bimodal distribution in mass ratio, q, for VLM dwarfs (Fig. 4): systems at small separations tend to have $q \sim 1$, while in wide binaries $q < 0.25$. The absence of lower-q systems at small separations might be a selection effect: as Fig. 3 illustrates, detecting faint companions becomes progressively more difficult with decreasing angular separation. On the other hand, several surveys offered the potential of detecting close systems with q as low as 0.5, yet none such are known. Thus, the observed tendency is probably a genuine property of low-mass binary systems. One can draw two conclusions from this result:

- First, this argues strongly against binary formation scenarios which envisage association of components drawn at random from the systemic mass function (indeed, there is essentially no evidence for such simplistic scenarios).
- Second, unrecognised binaries have been suggested as a possible source of bias in determining $\Psi(M)$. As originally pointed out by Reid [15], the bias is reduced substantially if significant fraction of the binaries are equal-luminosity / mass systems, as appears to be the case.

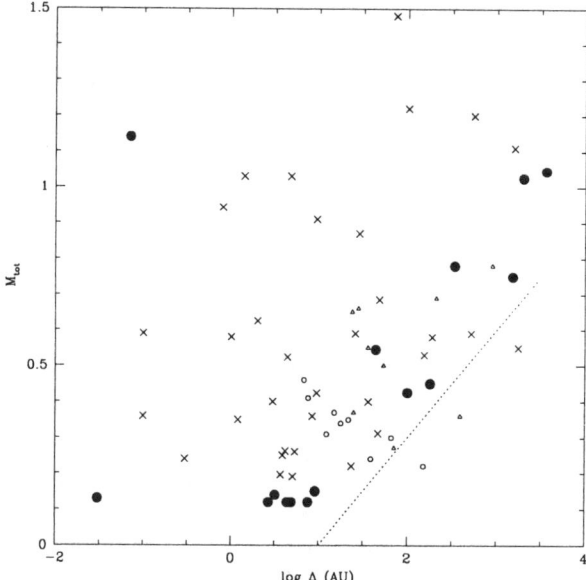

Fig. 5. Component separation as a function of total system mass: systems with brown dwarfs are plotted as solid points; nearby M dwarfs are plotted as crosses; Hyades M dwarfs as open circles; and field M dwarfs from RG97 as triangles.

The scarcity of wide VLM/VLM binaries is not a characteristics which suddenly becomes apparent amongst L dwarfs. Gizis & Reid [5] originally pointed out a relative scarcity of such systems in their observations of late-type Hyades M dwarfs; the L dwarf and Pleiades brown dwarf observations show that the deficit becomes more pronounced at lower masses. Indeed, plotting the separation of binary components against total system mass for a large sample of M dwarf and VLM dwarf binaries reveals evidence for a continuous trend, as illustrated in Fig. 5. The dotted line marked on that figure is hypothetical, rather than definitive, but suggests a rather steep (exponential?) decrease in the maximum likely separation of components with decreasing systemic mass. Based on this diagram, we predict that L dwarf/L dwarf binaries are *extremely* rare at separations exceeding 10 AU.

There are two possible explanations for the trend evident in Fig. 5:

- Low-mass binaries never form at separations exceeding 10 AU, perhaps reflecting decreasing size (or perhaps steeper density distributions) of lower-mass molecular cloud cores;
- Low-mass binaries form at intermediate and large separations, but have shorter lifetimes against gravitational disruption than higher-mass systems.

In the either context, the relatively steep dependence of the apparent cutoff on total system mass is somewhat surprising; in particular, the binding energy of a binary. system is dependent on the ratio $\frac{M_{tot}}{a}$, where a is the semi-major axis, so one might expect a significantly shallower gradient in a dynamically-defined cutoff radius. Further observations are clearly required to both confirm the reality of this effect and map out any age dependence.

4 Conclusions

Preliminary results from surveys searching for binary systems amongst L dwarfs discovered by the 2MASS, DENIS and SDSS surveys suggest both a preference for components with near-equal mass and a significant deficit of systems with separations exceeding 10 AU. Extending analysis to low-mass binaries reveals the presence of a trend in the (M_{tot}, Δ) plane, with an apparent cutoff in the maximum separation which decreases in a quasi-lognormal fashion with decreasing M_{tot}. It remains unclear whether this phenomenon reflects nature - such systems fail to form - or nurture - systems form, but are disrupted. High-resolution near-infrared observations of low-mass stars and brown dwarfs in young ($\tau < 10$ Myr) star clusters and millimetre data for protostellar cores in molecular clouds are needed to address this issue in more detail, although the substantial distances (>150 pc.) of suitable objects may prove a hindrance to acquiring adequate statistics in the immediate future.

References

1. E.E. Becklin, B. Zuckerman: Nature, **336**, 656 (1988)
2. X. Delfosse, C.G. Tinney, T. Forveille et al.: A&A, **327**, L25
3. A. Duquennoy, M. Mayor: A&A, **248**, 485 (1991)
4. D.A. Fischer, G.W. Marcy: ApJ, **396**, 178 (1992)
5. J.E. Gizis, I.N. Reid: AJ, **110**, 1248 (1995)
6. J.D. Kirkpatrick, I.N. Reid. J. Liebert et al.: ApJ, **519**, 802 (1999)
7. J.D. Kirkpatrick, I.N. Reid. J. Liebert et al.: AJ, **120**, 447 (2000)
8. D. Koerner, J.D. Kirkpatrick, M.W. McElwain, N.R. Bonaventura: **ApJ**, 526, L25 (1999)
9. S.K. Leggett, T.R. Geballe, X. Fan et al.: AJ, **536**, L35
10. E.L. Martín, W. Brandner, G. Basri: Science, 283, 1718 (1999)
11. E.L. Martín, W. Brandner, J. Bouvier et al.: ApJ, **543**, 299 (2000)
12. T. Nakajima, B.R. Oppenheimer, S.R. Kulkarni et al.: Nature, **378**, 463 (1995)
13. E. Öpik: Publ. Obs. Astron. Univ. Tartu, **25**, 6 (1924)
14. R. Rebolo, E.L. Martín, A. Magazzu: ApJ., **389**, L83 (1992)
15. I.N. Reid: AJ, **102**, 1428 (1991)
16. I.N. Reid, J.E. Gizis: AJ, **113**, 2246 (1997)
17. I.N. Reid, J.D. Kirkpatrick, J. Liebert et al.: ApJ, **521**, 613 (1999)
18. I.N. Reid, J.E. Gizis, J.D. Kirkpatrick, D.W. Koerner: AJ, **121**, 489 (2000)

The Second Guide Star Catalogue and Cool Stars

R.L. Smart[1], D. Carollo[1], M.G. Lattanzi[1], B. McLean[2], A. Spagna[1]

[1] Osservatorio Astronomico di Torino, Italy
[2] Space Telescope Science Institute, USA

Abstract. We discuss how the Second Guide Star Catalogue, GSC-II, will use cool stars to constrain the faint end of the luminosity function, the white dwarf contribution to dark matter in the halo and the nearby stellar population. We also assess the potential of GSC-II for research on the new ultracool L and T dwarfs.

1 The Second Guide Star Catalogue

GSC-II [6] will contain positions, proper motions, colours and classifications for over 1 billion objects. This is a 50 fold increase on the number of objects in the first catalogue and a wealth of new parameters. For most of the sky the catalogue will be complete to 18.5 in V, lacking completeness only in the most crowded regions of the plane. For declination greater than -30 and in the galactic plane there will be magnitudes in at least 4 bandpasses and 3 in the rest of the sky. The classification is estimated to be 90% accurate to the 18.5 limit and the photometry, calibrated individually for each plate, accurate to 0.2 magnitudes.

With this much data, over 4 terabytes, a simplified export version will be released while the main dataset will be managed on a sophisticated database system at STScI. The increased number of observed object parameters make this catalogue a powerful scientific tool. Below we highlight some of the projects that will be using and looking at cool stars, primarily M,K and white dwarfs, finally we consider if the newly discovered ultracool L and T dwarfs will appear in the GSC-II.

2 Selection of Cool Stars

In theory the selection of cool stars in GSC-II is quite simple. They are faint and have high proper motions. In practise it is difficult to do in any automatic way because the radius of matching between plates must be set so high that many objects are mismatched and given high proper motions. Therefore in all but the best cases most high proper motion detections require visual confirmation.

For much of the detection we use only POSSII plates that have a epoch difference of 1-10 years. Therefore with a search radius of 10" we can expect to find stars with proper motions $> 1.0"$/yr. We use the following restrictions to limit the number of mismatches in the generated sample which are all visually confirmed using both POSS II and POSS I images.

120 R.L. Smart et al.

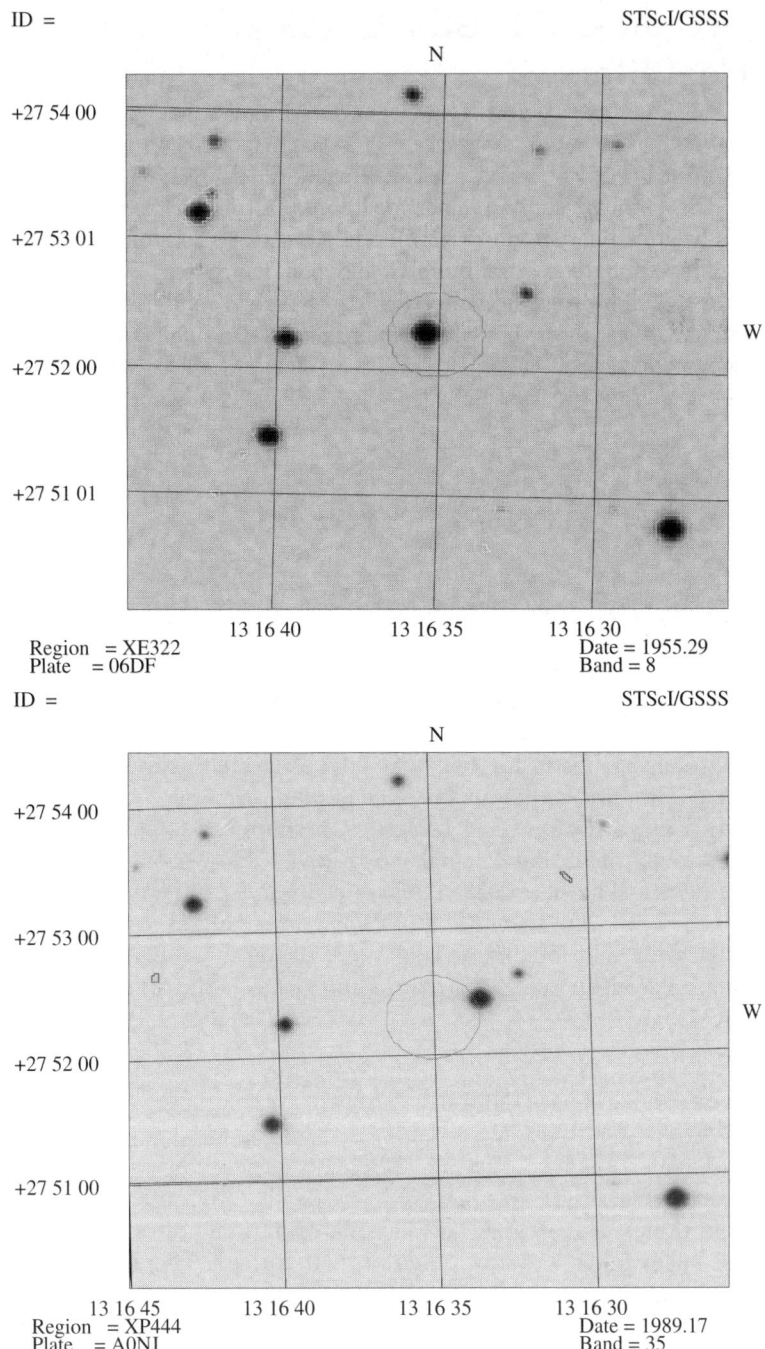

Fig. 1. A nearby red star with an evident proper motion shown here in the POSSII and POSSI red plates. A 10" circle is drawn for reference.

- object must be found on all three POSSII plates
- proper motion errors must be less than 3 σ
- overall classification must be stellar

Once we have these objects the colour in combination with the magnitude gives a good indication of what type of object it is. Most objects are either red dwarfs or disk white dwarfs, though we may also find halo white dwarfs, neutron stars or even bright L-dwarfs are expected to be visible on the red surveys.

3 The Faint End of the Local Luminosity Function

The majority of stars in the Galaxy are M-dwarfs and the luminosity function of these objects is therefore important for many studies, gravitational lensing, star formations, initial mass functions, baryonic dark matter and for comparison of our Galaxy with globular clusters. The present status of the local luminosity function knowledge is summarized in Kroupa et al (1993 [5]) and more recently in Gizis et al (2000 [2]) the error bars after $M_v < 10$ begin to reach 25%. This is due to the small number of stars known at these magnitudes. This determination disagrees with the photometric luminosity function using deep magnitude limited surveys for $M_v < 13$ for unknown reasons.

Using GSC-II we can increase the number of stars per bin in the range of interest and thus reduce the errors. To get the best results we require a large area of sky, say 10,000 square degrees, but with even 1000 square degrees we estimate we will have 10 stars at the faintest magnitude bin (M_v=17) and can reduce the errors by a factor of 2. We proceed as follows: find all stars with a proper motion between 0.04 ad 2.5 "/yr, using V-I colours determine an absolute magnitude, the calibration is shown in Fig. 2 for all low mass stars with trigonometric parallaxes. Red giants are eliminated by the low proper motion cut. For this work we will calibrate the GSC-II colours instead of working in the standard systems. Work in the NGP has begun for this study.

4 Searching for Dark Matter

That the missing mass could be made up of baryonic matter in the form of low mass stars was the impetus behind the MACHO, OGLE and EROS programs. Since these programs begun they have basically ruled out the possibility that objects between $10^{-7} - 10^{-2} M_\odot$ can make a significant contribution but they did find objects whose mass was indicated to be around the $0.5 M_\odot$ range, the mass of white dwarfs. Since this indication much work has gone into searching for halo white dwarfs in fields ranging from the 1'x1' HDF pointed field [3] to all sky search with the Luyten high proper motion catalogues. With the GSC-II we can check the contamination and completeness of the Luyten catalogues for measuring the significance of that work and we can also go fainter for selected regions. When you consider that the white dwarf contribution ranges from 1% [1] to 50% [7] any constraints will be useful. Indeed as many of the arguments

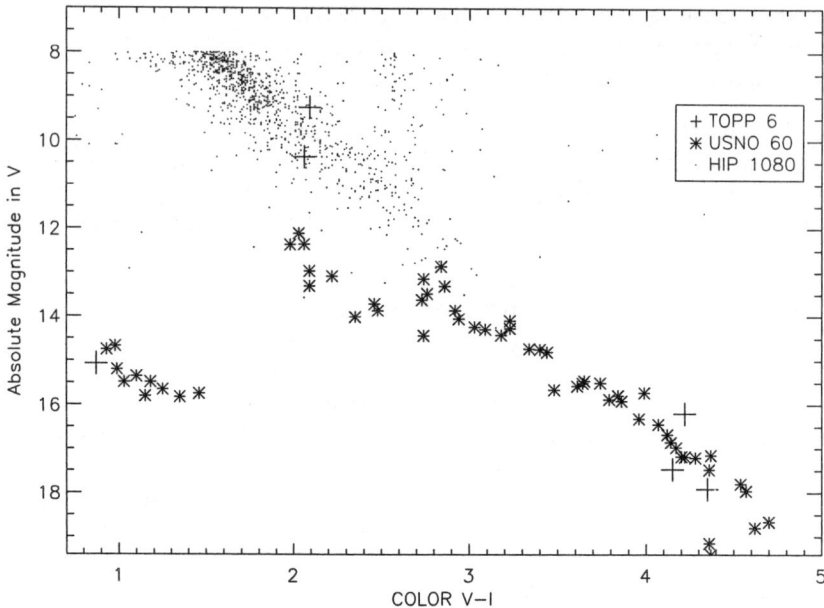

Fig. 2. Relation of Absolute Magnitude with $V - I$ colour for stars from Hipparcos, USNO [8], TOPP [9] programs. For stars with high proper motions the relation is almost unique.

have implications in other areas (i.e. extra galactic background light, the current ratio of carbon to hydrogen and the detection of gamma rays) even a limit of 10% would be interesting.

In the GSC-II program we use the above procedure to find the candidate stars, figure 1 is an example of one such candidate. The halo white dwarfs overlap with the brighter M dwarfs and disk white dwarfs in colour so we must use other information to distingush these populations. The proper motion, while quite indicative because of the halo white dwarfs expected high relative velocities, cannot be used alone, and we combine this information with the apprent magnitude in the reduced proper motion diagram (Fig. 3). We are planning to use spectroscopic observations for final confirmation, to determine the temperature of the object and to search for suppression of the infrared emission - a theoretically predicted indicator of cool white dwarfs.

5 L and T Dwarfs in the GSC-II

Kirkpatrick et al (2000) [4] in one fell swoop doubled the number of known L-dwarfs in this report on the discovery of 67 L-dwarfs in the 2MASS survey. Here they used the POSSII images to estimate the $R - K_s$ colour or when non detected in the POSSII at least put an upper limit on it. Gizis et al (2000) followed this work up with proper motion determinations of seven L-dwarfs using

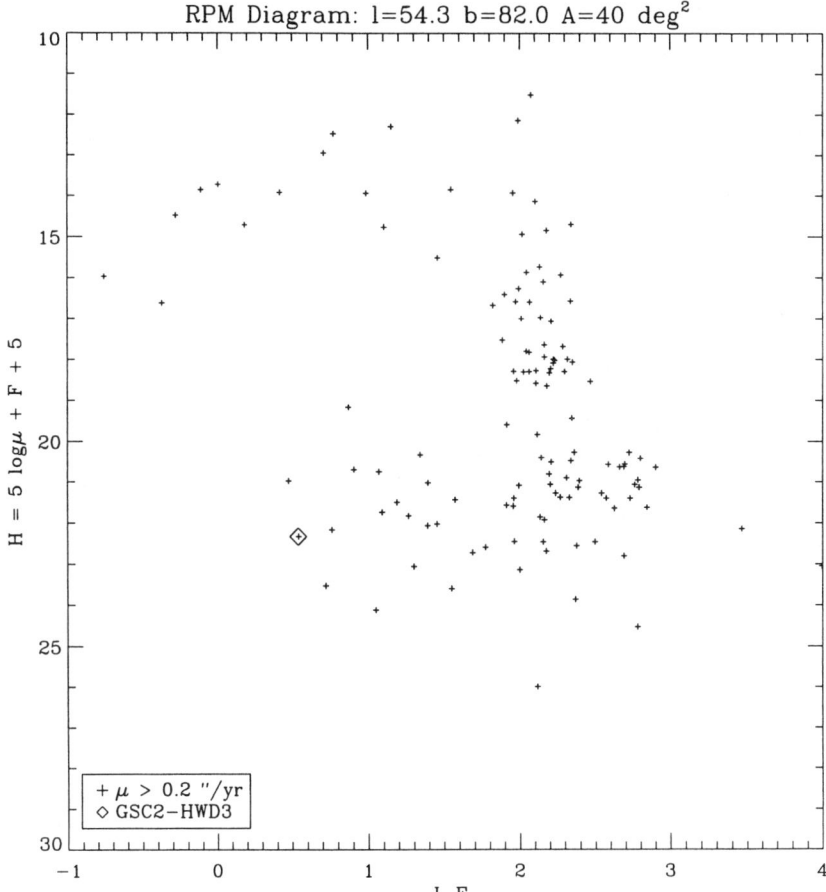

Fig. 3. Reduced proper motion diagram for a region near the north galactic pole for objects with proper motions above 0.2 arcseconds per year selected from the GSC-II. Stars with large RPMs, and blue colours are Halo WD candidates (e.g. GSC-II-HWD-3)

both POSSI, II and 2MASS positions. This paper was more concerned with the faint end of the luminosity function and the sample studied was mostly cool M-Dwarfs. Both of these works show the value of a catalogue such as GSC-II in the study of L-Dwarfs. As a result of their very red colours 2MASS and SDSS will remain the prime surveys for the discovery of these objects but for an accurate proper motion determination and for a rationalization of the discovery list the GSC-II will be of immense value.

References

1. K. Freese, B. Fields & D. Graff, in *Dark matter in Astrophysics and Particle Physics*, 397 (1999)
2. J.E. Gizis, D.G. Monet, I.N. Reid, J.D. Kirkpatrick, J. Liebert & R.J. Williams, AJ, **120**, 1085 (2000)
3. R.A. Ibata, H.B. Richer, R.L. Gilliland, & D. Scott, ApJ **524**, L95 (1999)
4. J.D. Kirkpatrick, I.N. Reid, J. Liebert, J.E. Gizis, A.J. Burgasser, D.G. Monet, C.C. Dahn, B. Nelson & R.J. Williams, AJ, **120**, 447 (2000)
5. P. Kroupa, C.A. Tout & G. Gilmore, MNRAS, **262**, 545 (1993)
6. B.M. Lasker, G.R. Greene, M.J. Lattanzi, B.J.McLean & A. Volpicelli, in *Astrophysics and Algorithms*, E3 (1998)
7. R.A. Mendez & D. Minniti, ApJ, **529**, 911 (2000)
8. D.G. Monet, C.C. Dahn, F.J. Vrba, H.C. Harris, J.R. Pier, C.B. Luginbuhl & H.D. Ables, ApJ, **103**, 638 (1992)
9. R.L. Smart, B. Bucciarelli, R. Casalegno, G. Chiumiento, F. Morale, M.G. Lattanzi, L. Lanteri, G. Massone, R. Morbidelli, F. Porcu & F. Racioppi, in R. Pallavicini and A. Dupree (eds.), *9th Cambridge Workshop, Cool Stars, Stellar Systems, and the Sun*, Astronomical Society of the Pacific (1995)

Low-Luminosity Companions to Nearby Stars: Status of the 2MASS Data Search

J.D. Kirkpatrick[1], J.E. Gizis[1], A.J. Burgasser[2], J.C. Wilson[3], C.C. Dahn[4], D.G. Monet[4], I.N Reid[5], and J. Liebert[6]

[1] Infrared Processing and Analysis Center, M/S 100-22,
California Institute of Technology, Pasadena, CA 91125, USA
[2] Division of Physics, M/S 103-33, California Institute of Technology,
Pasadena, CA 91125, USA
[3] Space Sciences, Cornell University, Ithaca, NY 14853, USA
[4] US Naval Observatory, Flagstaff Station, P.O. Box 1149, Flagstaff, AZ 86002, USA
[5] Department of Physics and Astronomy, University of Pennsylvania,
209 S. 33rd Street, Philadelphia, PA 19104, USA
[6] Steward Observatory, University of Arizona, 933 North Cherry Avenue,
Tucson, AZ 85721, USA

Abstract. We present 2MASS discoveries of four L and T dwarf companions to nearby stars. Using spectral types of the companions and age estimates of the primaries, we estimate masses for these discoveries and compare them to four other L and T dwarf companions from the literature. Temperatures range from ~2000 to ~750 K, ages range from ~0.2 to ~3 Gyr, and masses range from ~0.035 to ~0.075 $M_\odot$.

1 Introduction

Not since the pioneering work of [17] have vast numbers of nearby stars been searched for companions at wide separations, distances of a few hundred AU or more from the parent star. Characterization of such systems is necessary for a more complete census of the Solar Neighborhood. Specifically for brown dwarf research, there is another benefit: The discovery of L and T dwarf companions to age-datable stars allows us to estimate masses for the companions – something that cannot be done for L and T dwarfs in isolation.

We present preliminary work using the Two Micron All Sky Survey (2MASS) JHK_s data to search for previously unrecognized companions to nearby stars. Far optical and/or near-infrared spectra of these companions are shown and spectral classes, ranging from early-L through T, assigned. Using age estimates for the primary stars and temperatures inferred from the companions' spectral types, we estimate masses for each companion using theoretical work and compare these to several other L and T dwarf companions discovered by other groups.

2 Widely Separated L and T Dwarf Companions: 2MASS Discoveries

2.1 GJ 1048B: An Early-L Dwarf at 21.3 pc

The existence of a "bright", red ($K_s = 12.32$, $J - K_s = 1.35$) object only 11.9″ away (separation ~250 AU) from the K0 V star GJ 1048A (Fig. 1a), suggests that the two are physically associated. The spectrum of the red object (Fig. 1b) shows it to be an L1 dwarf, thus implying a distance similar to that of GJ 1048A and further suggesting that the two objects are associated. GJ 1048A has an X-ray luminosity implying an age older than the typical K dwarf in the Pleiades (age ~125 Myr) and more like a Hyad K dwarf (age ~625 Myr). It has, however, an X-ray luminosity brighter than 80% of the K dwarf disk population.

Based on these X-ray activity indicators, we estimate an age of 0.6-2 Gyr. For a more detailed discussion, see [5].

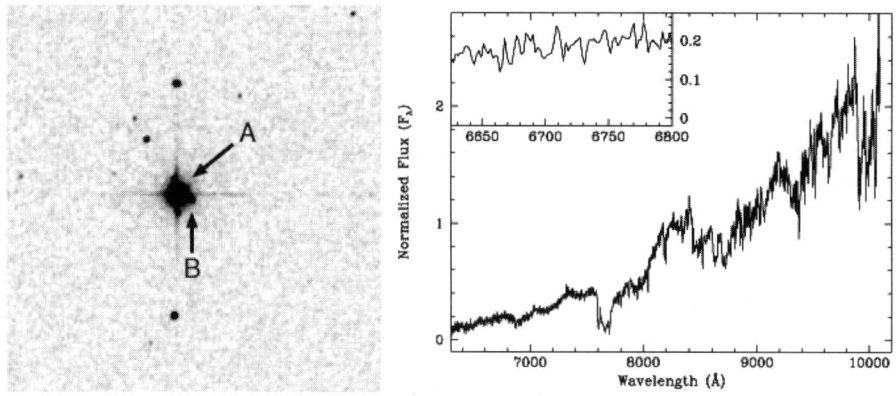

Fig. 1. (a) A 5′×5′ 2MASS K_s image of the GJ 1048AB system. (b) A far red spectrum of GJ 1048B, an L1 dwarf. Inset shows the area around 6708 Å.

2.2 Gl 417B: A Mid-L Dwarf at 21.7 pc

Confirmed via common proper motion to be a 90″-wide (separation ~2000 AU) companion to the G0 V star Gl 417A (Fig. 2a), this object has a spectral type of L4.5 V (Fig. 2b). The X-ray luminosity, rotational period, chromospheric variability, and lithium abundance suggest an age between 80 and 400 Myr for the G dwarf primary. A comparison of the primary's space motion with other stars shows that it is probably a member of the Local Association; a conclusion that is bolstered by the indicators listed above. The Local Association has an age of substantially less than 300 Myr [7].

Our age estimate for the Gl 417AB system is thus 80-300 Myr. For a more detailed discussion see [8].

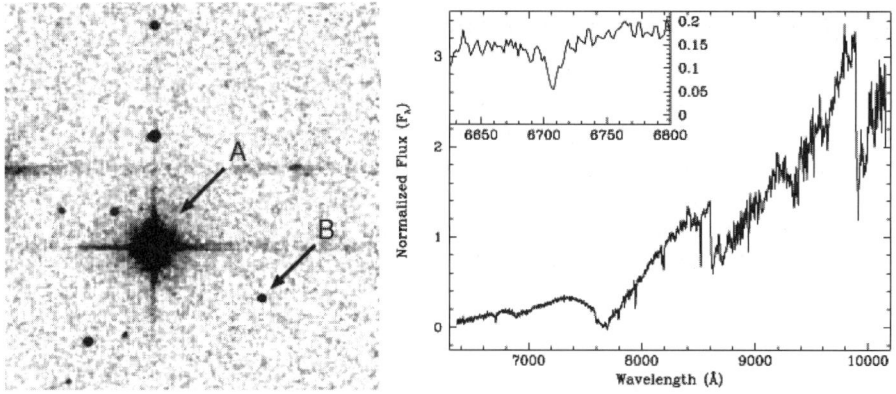

Fig. 2. (a) A $5' \times 5'$ 2MASS K_s image of the Gl 417AB system. (b) A far red spectrum of Gl 417B, an L4.5 dwarf. Inset shows the area around 6708 Å.

2.3 Gl 584C: A Late-L Dwarf at 18.6 pc

Confirmed via common proper motion to be a 194″-wide (separation ~3600 AU) companion to the G1 V + G3 V double Gl 584AB (Fig. 3a), this object has a spectral type of L8 V (Fig. 3b). The X-ray luminosity, rotational period, chromospheric variability, and lithium abundance suggest an age between 1 and 2.5 Gyr for the primary. Although the space motions of this G dwarf pair have caused it to be suspected as a member of the Ursa Major Moving Group (age of ~300 Myr), the preceeding diagnostics clearly disprove this [16].

Our age estimate for the Gl 584ABC system is thus 1-2.5 Gyr. For a more detailed discussion see [8].

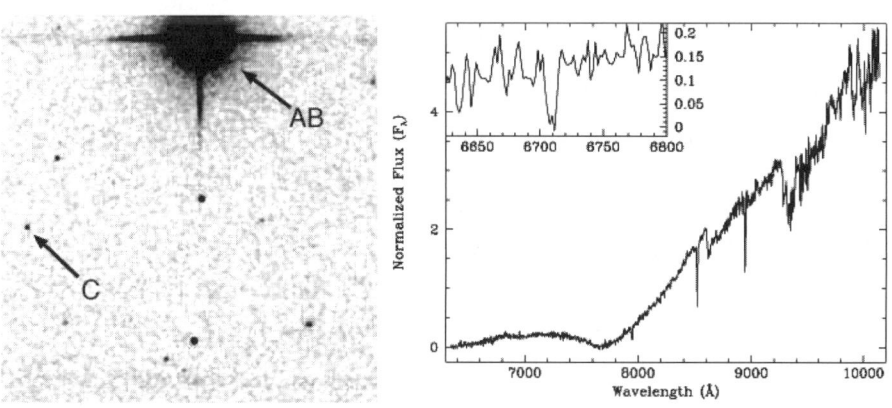

Fig. 3. (a) A $5' \times 5'$ 2MASS K_s image of the Gl 584ABC system. (b) A far red spectrum of Gl 584C, an L8 dwarf. Inset shows the area around 6708 Å.

2.4 Gl 570D: A T Dwarf at 5.9 pc

Confirmed via common proper motion to be a 258″-wide (separation ∼1500 AU) companion to the K4 V + M1.5 V + M3 V triple Gl 570ABC (Fig. 4a), this object has a T dwarf spectral type (Fig. 4b). A lack of activity in the close M dwarf pair along with measures of the X-ray and UV luminosities of the triple suggests an age older than 2 Gyr. Kinematics also suggests the system is a member of the Galactic disk, supporting an age younger than 10 Gyr.

Our age estimate for the Gl 570ABCD system is thus 2-10 Gyr. For a more detailed discussion, see [3].

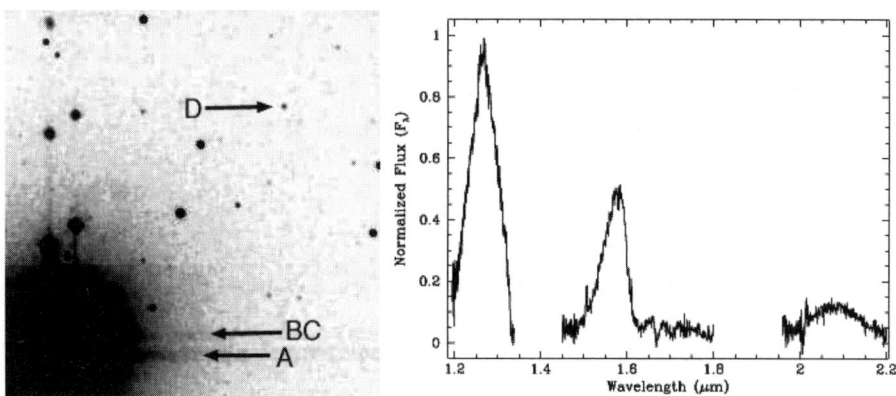

Fig. 4. (a) A 5′×5′ 2MASS J image of the Gl 570ABCD system. (b) A near-infrared spectrum of Gl 570D, a T dwarf.

3 Other L and T Dwarf Companions: Discoveries by Other Groups

The primaries of these systems are M dwarfs or white dwarfs for which accurate ages are harder to derive.

3.1 G 196-3B: An Early-L Dwarf at ∼20 pc

Discovered by [14], this object has been confirmed via common proper motion to be the 16.2″-wide (separation ∼300 AU) companion to the ∼M2.5 dwarf G 196-3A (Fig. 5a). Our far red spectrum (Fig. 5b) has a spectral type of L2 V [8]. Activity in the M dwarf sets a rough age of 60-300 Myr for the system [14].

3.2 GD 165B: A Mid-L Dwarf at 31.5 pc

Discovered by [2], this object has been confirmed via common proper motion [19] to be the 3.7″-wide (separation ∼120 AU) companion to the DA4 white dwarf

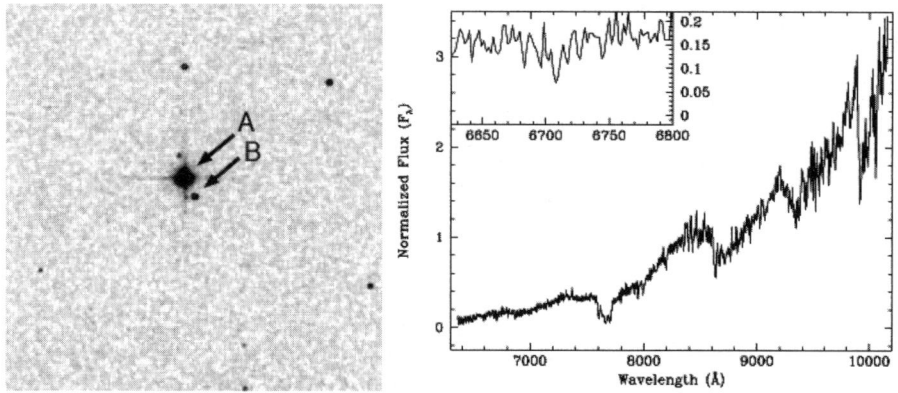

Fig. 5. (a) A 5'×5' 2MASS K_s image of the G 196-3AB system. (b) A far red spectrum of G 196-3B, an L2 dwarf. Inset shows the area around 6708 Å.

GD 165A (Fig. 6a). Our far red spectrum (Fig. 6b) has a spectral type of L4 V. The white dwarf primary sets a coarse age of 1.2-5.5 Gyr for the system [9].

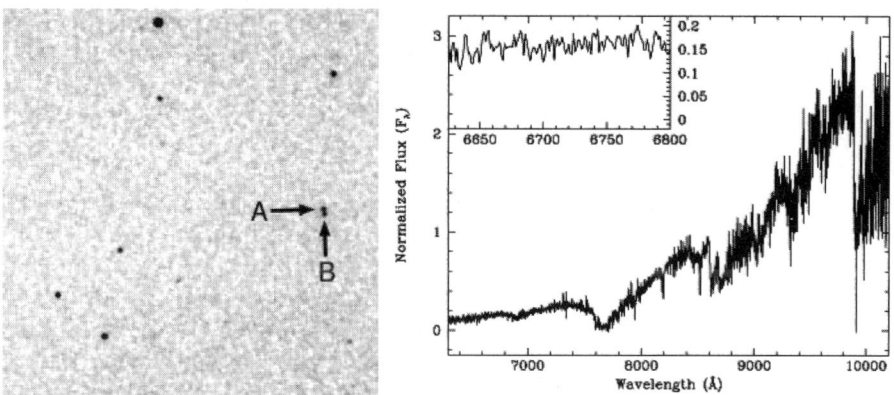

Fig. 6. (a) A 5'×5' 2MASS H image of the GD 165AB system. (b) A far red spectrum of GD 165B, an L4 dwarf. Inset shows the area around 6708 Å.

3.3 GJ 1001B: A Mid-L Dwarf at 9.6 pc

Discovered by [6] and also known as LHS 102B, this object has been confirmed via common proper motion to be the 18.6"-wide (separation ∼180 AU) companion to the M3.5 dwarf GJ 1001A (Fig. 7a). Our far red spectrum (Fig. 7b) has a spectral type of L5 V [8]. Very crude age estimates place the primary at an age older than 1 Gyr [6].

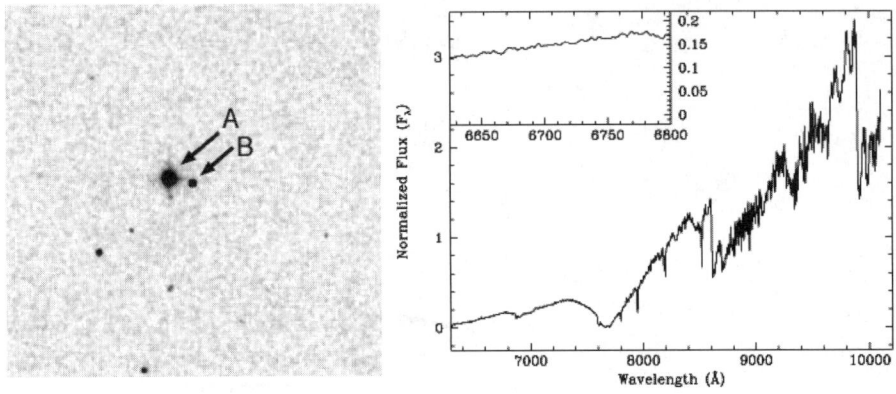

Fig. 7. (a) A 5'×5' 2MASS K_s image of the GJ 1001AB system. (b) A far red spectrum of GJ 1001B, an L5 dwarf. Inset shows the area around 6708 Å.

3.4 Gl 229B: A T Dwarf at 5.8 pc

Discovered by [11], this object has been confirmed via common proper motion to be the 7.7''-wide (separation ~45 AU) companion to the M1 dwarf Gl 229A. The near-infrared spectrum [12] shows methane bands, the hallmark of spectral class T. Activity measurements for the M dwarf suggest an age of 0.6-5 Gyr [11].

4 Mass Estimates for Each Companion

Using age estimates of the eight nearby systems and using temperature estimates derived from each companion spectrum, we can estimate masses for each companion using theoretical evolutionary tracks. As summarized in Table 1, a range of temperatures has been assigned to each spectral class where estimates for the L dwarfs are bracketted by the "warm" scale favored by [1] and the "cool" scale favored by [10] and [15]. Using the age and temperature estimates listed in Table 1, we can plot the locus of each companion on the theoretical HR Diagram of [4], as shown in Fig. 8, and hence derive a mass estimate for each. These mass estimates are summarized in Table 1.

Finally, we can use the presence or absence of 6708-Å Li I absorption in the L dwarf companions[1] as an independent check of the masses. As originally pointed out by [13], lithium is a powerful tool in the study of brown dwarfs. Lithium is not produced as a stable byproduct of the proton-proton chain, so any object that is fully convective – and all low mass stars and brown dwarfs spend at least part of their lives in fully convective states – will, if hot enough to burn lithium, rapidly burn its entire primordial stock since there is no means for replenishing

---

[1] For the two T dwarfs discussed here, Gl 570D does not have sufficient signal at 6708 Å to judge the presence or absence of lithium, and Gl 229B has not been observed spectroscopically at this wavelength.

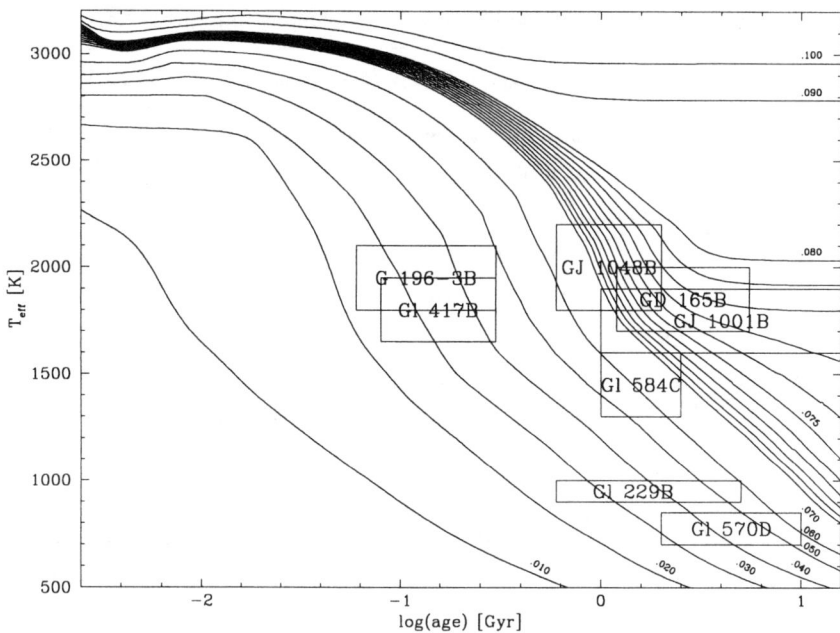

Fig. 8. Theoretical evolutionary tracks from [4] for low-mass stars and brown dwarfs. Tracks are labelled with mass in solar masses ($M_\odot$). Plotted as boxes are the age and temperature ranges for each of the eight L and T dwarfs in Table 1. Note that the dwarfs G 196-3B, Gl 417B, Gl 229B, and Gl 570D may be thought of as a ~0.040 $M_\odot$ brown dwarf seen at four different stages in its life.

it. Lithium burns at temperatures exceeding $\sim 2.5 \times 10^6$ K; theoretical models show that only objects above ~ 0.060 $M_\odot$ have maximum central temperatures in excess of this value. This means that the presence of lithium in a low-mass object indicates that its mass is below ~ 0.060 $M_\odot$.

If we check Table 1, we find that the L dwarf companions without Li I absorption have mass estimates >0.060 $M_\odot$ and those with Li I absorption have estimates <0.060 $M_\odot$, in agreement with expectations. Only for Gl 584C does lithium provide additional constraints on the mass and in this case the presence of lithium favors a mass in the lower half of our estimated range.

5 Conclusions

We are continuing this program to search the 2MASS database for possible companions around *all* known stars and brown dwarfs within 25 parsecs of the Sun. Another three candidate companions found by 2MASS and ranging in type from late-M through late-L will be reported in [18]. Continued searches will give us a more accurate accounting of the nearby census, allow us to build a library

Table 1. Summary of L and T Dwarf Companions to Nearby Stars

Object	Type	Age (Gyr)	Temp. (K)	Mass ($M_\odot$)	Li I EW
GJ 1048B	L1 V	0.6–2.0	1800–2200	0.070±0.010	<1 Å
G 196-3B	L2 V	0.06–0.3	1800–2100	0.035±0.015	6 Å
GD 165B	L4 V	1.2–5.5	1700–2000	0.075±0.005	<0.7 Å
Gl 417B	L4.5 V	0.08–0.3	1650–1950	0.035±0.015	11.5 Å
GJ 1001B	L5 V	>1.0	1600–1900	0.070±0.010	<0.2 Å
Gl 584C	L8 V	1.0–2.5	1300–1600	0.060±0.015	7.4 Å
Gl 229B	T	0.6–5.0	900–1000	0.040±0.020	–
Gl 570D	T	2.0–10.	700–850	0.045±0.020	–

of L and T dwarfs of known age (and metallicity) and inferred mass, and enable us to study the results of binary formation process for mass ratios far from unity.

References

1. G. Basri, S. Mohanty, F. Allard, P. H. Hauschildt, X. Delfosse, E. L. Martín, T. Forveille, & B. Goldman: ApJ **538**, 363 (2000)
2. E. E. Becklin, B. Zuckerman: Nature, **336**, 656 (1988)
3. A. J. Burgasser, et al.: ApJ, **531**, L57 (2000)
4. A. Burrows, et al.: ApJ, **491**, 856 (1997)
5. J. E. Gizis, J. D. Kirkpatrick, J. C. Wilson: AJ, submitted (2000)
6. B. Goldman, et al.: A&A, **351**, L5 (1999)
7. R. D. Jeffries: MNRAS, **273**, 559 (1995)
8. J. D. Kirkpatrick, et al.: in prep. (2000)
9. J. D. Kirkpatrick, F. Allard, T. Bida, B. Zuckerman, E. E. Becklin, G. Chabrier, I. Baraffe: ApJ, **519**, 834 (1999)
10. J. D. Kirkpatrick, et al.: AJ, **120**, 447 (2000)
11. T. Nakajima, B. R. Oppenheimer, S. R. Kulkarni, G. A. Golimowski, K. Matthews, S. T. Durrance: Nature, 378, 463 (1995)
12. B. R. Oppenheimer, S. R. Kulkarni, K. Matthews, T. Nakajima: Science, **270**, 1478 (1995)
13. R. Rebolo, E. L. Martín, A. Magazzù: ApJ, **389**, L83 (1992)
14. R. Rebolo, M. R. Zapatero Osorio, S. Madruga, V. J. S. Bejar, S. Arribas, J. Licandro: Science, **282**, 1309 (1998)
15. I. N. Reid, et al.: AJ, in press (2000)
16. D. R. Soderblom, M. Mayor: AJ, **105**, 226 (1993)
17. G. van Biesbroeck: AJ, **66**, 528 (1961)
18. J. C. Wilson, et al.: in prep. (2000)
19. B. Zuckerman, E. E. Becklin: ApJ, **386**, 260 (1992)

Part III

Spectral Classification

Introduction: The Spectral Types of the Ultracool Dwarfs

M.S. Bessell

The Research School of Astronomy and Astrophysics, IAS,
The Australian National University, Cotter Rd, ACT 2611, Australia

Abstract. The proposed L and T spectral classes to classify Ultra-Cool Dwarfs are well defined and extend the K and M sequence of the MK system to lower temperatures. They can be understood as a single temperature sequence by including dust condensation and segregation in addition to ionization and dissociation that underlies the MK system.

1 Introduction

As stated in this volume [3] the purpose of classification is to place objects that are similar in the same category and to place those that are dissimilar into different categories. The important observation was also made that the use of a unique spectral class that is invariant with time enables clear communication between scientists by providing a mental picture of a spectrum associated with that spectral type. Although these criteria remain most important for spectroscopists, the underlying physics of the stellar photosphere determines the long time usefulness of any classification system to a wider community of astronomers.

The brilliance of the MK two dimensional classification system was that its O-M sequence could be understood as a temperature sequence (with luminosity as the second dimension) by considering ionization and dissociation in a gaseous mixture. In addition, this 2-dimensional system naturally mapped onto the HR diagram of main sequence and post-main sequence stars and through that into stellar isochrones and evolutionary tracks.

The challenge for astronomers today is to devise, if possible, an extension of the O-M sequence that incorporates the L and T spectra of the ultra-cool dwarfs and which reflects the underlying chemical and physical equilibria of the stellar photosphere and again maps onto the HR diagram of main sequence, pre-main sequence stars and brown dwarfs.

Spectral classification, that was originally based on line ratios in the blue-visible region, has been extended to other wavelengths to take advantage of better detector sensitivities and higher stellar fluxes. Care is normally taken to maintain the uniqueness of the spectral type derived in the different spectral regions. More problematic has been the extension in the use of spectral types to label colours. Apart from a problem with semantics, correlating colours and spectra can be uncertain. Some colours and some spectral types correlate extremely well thus leading to the practice in the first place; however, in some spectral regions and for some colours there is poor correlation. Such disagreements can provide useful information about the stellar atmosphere but more often are used to argue whether a spectral type is right or wrong. For F G K and M stars colours such as B-V, V-I or V-K are excellent indicators of temperature but when the fluxes are affected by dust reddening or by metal deficiency or by atmospheric turbulence, such colours can be uncertain temperature indicators. In the late-M, L and T dwarfs, correlations between some colours and spectral types show large scatter leading to different temperature derivations and conflict in interpretation of the nature of the stars.

Over the past few years knowledge of the nature of the ultra-cool dwarfs spectra has rapidly matured. This is clearly evident from the papers presented at this conference, where spectra and colours for samples of UCDs were discussed and in addition, synthetic spectra and colours for different theoretical models were presented. Although some spectral-type purists may prefer spectral classification to be carried out without reference to theoretical modelling, most people regard the insight thus provided to be invaluable.

2 Discussion

2.1 Spectral Types

The nature of the MK classification types is that each class can be considered to be centered about a particular spectral feature or features that tend to dominate the spectrum and/or whose strengths change the most rapidly in blue-visual spectra. For example, the HeI lines in B stars, the HI lines in A stars and the TiO/VO bands in M stars. In the L dwarf sequence the dominant features are the strengthening wings of the NaI and KI lines. the weakening of the TiO and VO bands and the waxing and waning of the hydride bands. The T dwarfs spectra are dominated by absorption bands of H_2O and CH_4. In discrete classification schemes there are likely to be difficulties at the edges in deciding whether a star should be a late member of one class or an early member in another. This is alleviated in most spectral type divisions by not making use of all ten spectral-subclasses, as for example in the jump progressions A5, A7, F0, and K5, K7, M0. It appears that this could usefully be done with the late M and late L classes as well.

The one dimensional L spectral subclasses established by Martin et al. [5] and Kirkpatrick et al. [3] are based mostly on CrH, FeH, TiO and VO band ratios plus red pseudo-continuum colours. Kirkpatrick [3] [2] shows the relations between his subclasses and the strengths of the CrH band and the lines of RbI and CsI. Similar plots for CrH, FeH, TiO and VO are presented by Martin et al.[5] and the large scatter for the M8-L0 dwarfs highlights 1-D classification problems at the join. The Kirkpatrick and Martin subclasses are in agreement for L1-L3 but differ for later types, the Martin classes being earlier.

Stephens, Marley & Noll [6] have obtained near IR colours for a sample of L and T dwarfs in the near-IR K, L_s and L' bands. They claim that $K - L$

colours should be little affected by dust and will therefore be a good temperature index. Their $K - L'$ index plotted against Kirkpatrick spectral classes shows a reasonable correlation but for the L dwarfs the scatter is ± 2 divisions. For their sample of L5-L8 dwarfs they find little change in $K - L'$. The $K - L_s$ colour which is also a measure of the fundamental CH_4 band shows a sharp drop for the T dwarfs and increased scatter for the late L dwarfs.

2.2 Model Atmosphere Insights

Tsuji has long been a pioneer in the study of molecular opacities, molecular equilibria in the atmospheres of cool stars. He and his coworkers constructed 'dusty' and 'dust-segregated' models in efforts to match the spectra of the archetypal L and T dwarfs [8],[9] and to understand inconsistencies in the spectra of the late-M dwarfs [1]. In this proceedings we see the impressive results of the new unified model atmospheres which incorporate an active warm dust zone. The spectra of these models are a significant improvement on previous modelling and now match the observational colours and spectra of M, L, and T dwarfs extremely well. There are still some deficiencies, especially in the computed K band fluxes, but the overall impression given is that the basic chemistry and physics of cool star atmospheres are now understood. As a result of this work, Tsuji [7] states confidently that the sequence of spectral types M, L and T can be considered as a single temperature sequence by considering dust condensation and segregation in the stellar atmosphere in addition to ionization and dissociation that is considered in hotter stars.

3 Luminosity Extension to Spectral Types

Whilst Tsuji's unified models successfully fit the overall spectra and explain the temperature of transition between M-L and L-T, it is obvious that there is considerable scatter in relations between the spectral types and the far-red and near-IR colours of the late-M, L and T dwarfs. Some scatter can be expected from stars having different levels of chromospheric activity and/or metallicity. But as pointed out by Martin [4], more scatter is to be expected in a sample of stars because the existing spectral-types are based on spectral features, many of which change strength with effective gravity or luminosity. Martin suggests that analysis of a large sample of spectra should be able to extract both a temperature and a luminosity index from the spectral features and colours. Extensive modelling should also be able to predict such indices.

It is well worthwhile to attempt this luminosity extension and a hoped for reduction in the scatter between spectral-types and colours. But irrespective of the success of that endeavour, it would be useful to use fewer divisions in both the late M and the late L spectral-type groups to encompass the astrophysical scatter and recognise that precise spectral type subdivisions, implying precise temperature determinations, are not yet possible.

References

1. H.R.A. Jones & T. Tsuji, ApJ **480**, L39 (1997)
2. J.D. Kirkpatrick, this volume, (2001)
3. J.D. Kirkpatrick, F. Allard, T. Bida, B. Zuckerman, E.E. Becklin, G. Chabrier & I. Baraffe, ApJ, **519**, 834 (1999)
4. E.L. Martin, this volume (2001)
5. E.L. Martin, X. Delfosse, G. Basri, B. Goldman, T. Forveille & M.R. Zapatero Osorio, AJ, **118**, 2466 (1999b)
6. D.C. Stephens, M.S. Marley & K.S. Noll. this volume (2001)
7. T. Tsuji, this volume (2001)
8. T. Tsuji, K. Ohnaka, & W. Aoki, A&A **305**, L1, (1996)
9. T. Tsuji, K. Ohnaka, W. Aoki & J. Nakajima, A&A **308**, L29 (1996)

The Classification of L Dwarfs

J.D. Kirkpatrick

Infrared Processing and Analysis Center, M/S 100-22,
California Institute of Technology, Pasadena CA 91125, USA

"Astronomical spectroscopy is an almost magical technique. It amazes me still."
– *Carl Sagan (1980) [26]*

Abstract. To highlight the need for new spectral types, a brief review of the discovery of very low luminosity dwarfs – cooler than type M9.5 – is presented. The reasoning behind the choice of letters "L" and "T" is also given. Our philosophy of spectral classification is explained and we use this philosophy to construct an L dwarf typing scheme. The scheme, originally developed on only 25 spectra, has now been shown to work successfully on a sample of 92 L dwarfs.

1 Introduction

When John Ray and Carl Linneaus first began categorizing plants in the late 1600's and early 1700's, the methodology of modern scientific classification was born. Large collections were sorted empirically into natural groupings and sub-groupings and given names that could be easily remembered and discussed. In astronomy, it has been only in the last two hundred years that stars could be classified on anything other than an artificial system involving their brightnesses or perceived colours. With the application of stellar spectroscopy in 1814 by Joseph Fraunhofer, an individual star was transformed from just another point of light to a classifiable object worthy of detailed study. Slowly, techniques were developed to categorize large numbers of stars based on their spectra and in 1866 Angelo Secchi produced the first major work on stellar spectral classification. This scheme was refined over the next 35 years, and by 1901 stellar spectral types had evolved into the hot-to-cool OBAFGKM sequence recognized today [5].

One hundred years later, advanced detector technology and aggressive surveying of large areas of sky have enabled us to discover objects much cooler than M-type stars. Classification of these newest discoveries is the subject of this paper and several others in this volume.

2 Why Do We Need New Types?

The near-infrared colour-colour diagram of Fig. 1a depicts the bottom of the main sequence as we knew it in early 1997. Solid circles show M dwarfs, from early-M types at $(J-H, H-K_s) \approx (0.6, 0.2)$ to late-M dwarfs at $(J-H, H-K_s) \approx (0.7, 0.5)$ [15]. These were the lowest luminosity objects recognized with the exception of GD 165B at $(J-H, H-K_s) = (0.93, 0.65)$ [12] and Gl 229B at

$(J - H, H - K_s) = (-0.1, 0.0)$ [17]. To guide the eye, the locus of dwarfs earlier than type M is shown by the solid line; whereas, late-type giants are shown by the dashed line [3].

GD 165B (open circle in Fig. 1a), located 3.7″ away from the DA4 star GD 165A, had been discovered earlier [2] during a near-infrared imaging campaign designed to look for brown dwarfs around nearby white dwarfs. Common proper motion confirmed it to be physically associated with the A component [29], thus establishing a luminosity, $L \approx 1 \times 10^{-4} L_\odot$, much lower than any M dwarf. A red spectrum of the object [8] also showed it to lack familiar TiO bands that are the hallmark of spectral class M. A recently acquired spectrum of this object is shown in Fig. 2. GD 165B was a unique object, alone in its portion of colour space.

Even lower in luminosity was Gl 229B as illustrated by the starred symbol in Fig. 1a. It had been discovered [20] during a coronagraphic/adaptive optics campaign designed to find brown dwarf companions to relatively young nearby stars. Common proper motion with the M1 V star Gl 229A confirmed physical association, and a luminosity of $L \approx 6 \times 10^{-6} L_\odot$ was established – significantly cooler even than GD 165B. A near-infrared spectrum [22] showed that it looked nothing like either the late-M dwarfs or GD 165B. In fact, the reason for its blue near-infrared colours – it lies at the same place in Fig. 1a as A-type stars – is that it has deep methane bands absorbing much of the flux at H and K_s.

In a spectroscopic sense, these two objects are fundamentally different from M dwarfs and from one another. Hence, new spectral types to follow "M" appeared to be needed. The assignment of new spectral types began in earnest when, in 1997, field objects were found which bridged the gap between late-M dwarfs and GD 165B [9,6,25].

3 Choosing the Letters

Which letters are left for use as possible new spectral classes? As stated in [10] the best remaining letters are "H", "L", "T", and "Y" after previously-used and confusing letters are eliminated. Reasons for avoiding other letters of the alphabet are tabulated in [11]. If either "H" or "L" were used, it would be a letter originally used in but later dropped from the Harvard classification system [23,5]. Neither "T" nor "Y" has been previously used as a spectroscopic class.

As even more objects similar to GD 165B were being discovered between 1997 and 1998 by the Two Micron All Sky Survey (2MASS) [11], I held discussions with colleagues in the field and with members of IAU Commission 45 on Stellar Classification to address the need for new spectral types to follow "M". The vast majority felt that new designations separate from "M" were indeed warranted based on the new spectra being collected. Of the preceding four letter choices, "L" was the overwhelming favorite to follow "M". It was decided that beyond this, one would proceed alphabetically down the list of remaining letters. Following "L" would be "T", to designate spectra such as Gl 229B with methane absorption at 2.2μm. Following "T" would be "Y" if needed later [11].

The Classification of L Dwarfs 141

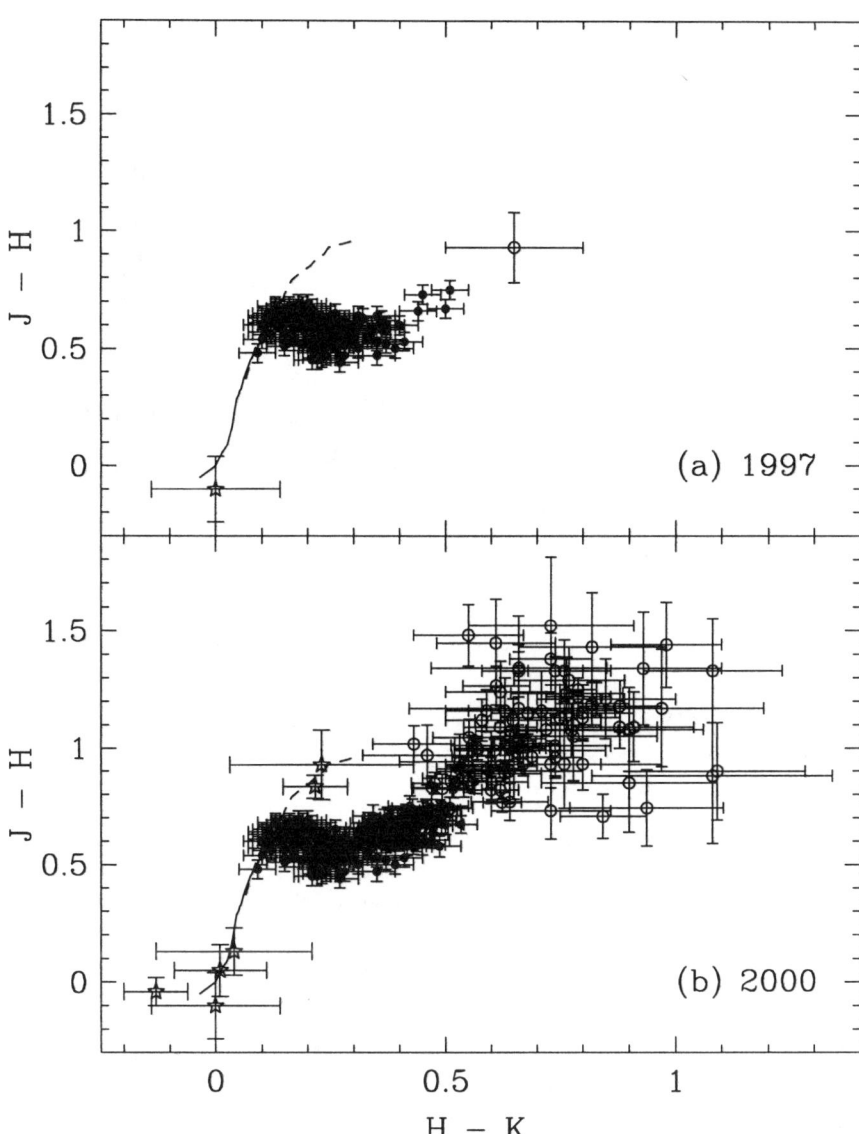

Fig. 1. (a) The near-infrared colour-colour plot of the bottom of the main sequence as it was known in early 1997. Solid points are M dwarfs, the open circle is GD 165B, and the open star is Gl 229B. (b) The same diagram as it is known today, with many more M dwarfs (solid circles), L dwarfs (open circles), and T dwarfs (open stars) now recognized. The locus of early-type dwarfs is shown in both panels by the solid line, and the locus of red giants is shown by the dashed line.

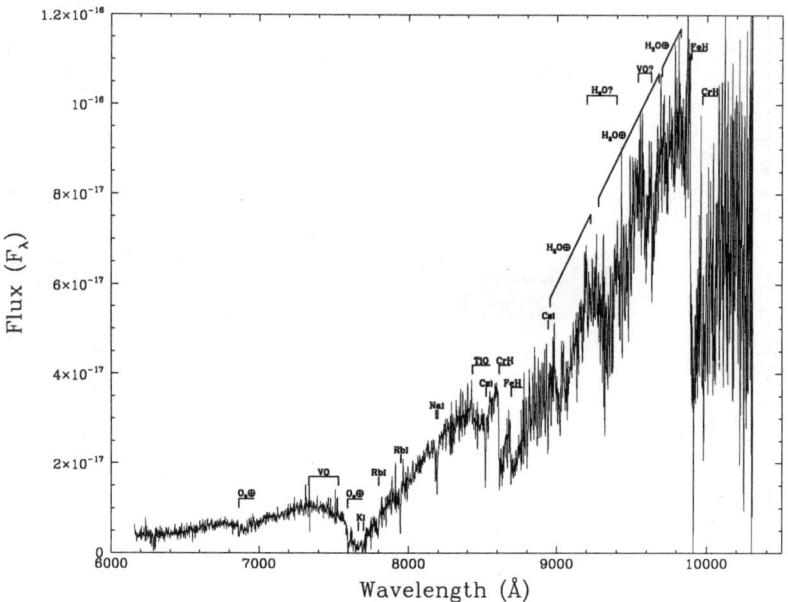

Fig. 2. A 6300-10000 Å spectrum of GD 165B with prominent line and band identifications marked [12].

Also, I asked Phil Keenan, one of the fathers of the MK Classification System, for his opinion. Although he did not have a strong preference for the choice of letters, he was very interested about the categorization of the spectra. In his response he wrote, *"Thank you for letting me see these fascinating spectra. I'll look forward to your further results. Perhaps when your classification is complete, I can add stars of your class to our catalog of spectral standards!"*

The first sizable set of L dwarfs – 25 in all – was published in 1999 [11]. The classification scheme of that paper and its extension to objects discovered since then are discussed below. T dwarfs, of which 23 examples are known at the time of this writing, are the subject of a classification paper presented elsewhere in this volume [4].

4 Establishing the L Dwarf Classification System

4.1 Philosophy

In defining individual subclasses for L dwarfs, there are two questions that need to be addressed: (1) Over what wavelength range and at what resolution should the L-dwarf classification system be defined? (2) What should be the basis of classification?

To answer question 1, we must first recognize the fact that it is necessary to tie L dwarf classifications to the pre-existing classification for M dwarfs. We have thus chosen a very similar spectral region and resolution to that used in [7] to define M dwarf classes. Namely, we have selected the far red spectral region[1] between 6300 and 10000 Å at a resolution of 9 Å. As shown in Fig. 2, this spectral region includes bands of TiO and VO which ma be used to differentiate between the latest M dwarfs and the earliest L dwarfs. The region also includes one set each of resonance lines from the neutral alkali metals Li, Na, K, Rb, and Cs as well as two bands each of CrH and FeH.

To answer question 2, we need only ask ourselves why we are classifying these objects in the first place. The purpose of classification is to place objects that are similar into the same category and to place those that are dissimilar into different categories. Use of a classification label makes for easy communication between scientists. Use of a spectral class such as "M7" gives astronomers a mental picture of the type of spectrum (strong TiO bands and weak VO bands in the optical) being discussed. Hence, the correct way to assign spectral types is through spectral morphology. As W. W. Morgan – the other father of the MK Classification System – wrote, standards of reference *"do not depend on values of any specific line intensities or ratio of intensities; they have come to be defined by the appearance of the totality of lines, blends, and bands"* [19]. Spectral typing is merely the first step in trying to extract the physics governing these spectra. Note, however, that a physical picture is *not* used in assigning types nor would we want to use any physical model as a basis for the classification. Spectral types should remain unchanged with time even if our understanding of the physical mechanisms shaping the spectra does not. An object classified as L7 today should also be classified as L7 ten or a hundred years from now. It is the immutability of the types that makes scientific dialogue comprehensible.

4.2 The Qualitative Ordering

In the latter half of 1997, we began selecting L dwarf candidates from the first scans acquired by the all-sky survey of 2MASS. The photometric selection criteria used to select candidate targets is addressed elsewhere [11]. By January of 1998, the 2MASS Rare Object Team[2] had confirmed twenty of these to be objects later than type M9.5 using the Low-Resolution Imaging Spectrograph [21] at the 10m W. M. Keck Observatory on Mauna Kea, Hawaii. These 6300-10000 Å spectra, along with spectra we acquired of five similar objects from the literature, gave us a collection of twenty-five L dwarfs upon which to base a classification scheme.

Armed with print-outs of the twenty-five L dwarf spectra along with several late-M dwarfs, I spread the pages across my living room floor and attempted

---

[1] If instrumental restrictions prohibit the acquisition of spectra identical to this, a smaller wavelength coverage at identical resolution will suffice. In this case, coverage between 7000 and 9000 Å is adequate for classification purposes.

[2] At that time, the Rare Object Team was comprised of Jim Liebert, Dave Monet, Roc Cutri, Neill Reid, Conard Dahn, John Gizis, Brant Nelson, and myself.

to construct a sequence. I would order the spectra starting with those most resembling late-M dwarfs and ending with the ones most different. On the back of each print-out I would jot down each spectrum's place in the ordering, reshuffle the pages and again order them from scratch a week later. After the first three weeks, I had converged to the same ordering. Binning the spectra into subtypes was done the same way; selecting a representative spectrum (i.e., spectral standard) for each subtype x so that differences between it and the standards designating subtypes x-1 and x+1 were obvious. Using this method I found that the spectra fell naturally into nine bins, thereafter denoted as L0, L1, L2, ..., L8. This ordering of the spectral standards is shown in Fig. 3.

4.3 The Quantitative Ordering

With this by-eye ordering, quantitative criteria for use as classification diagnostics were then established. Figure 4 shows detailed regions of the spectra shown in Fig. 3. Note the slow disappearance of TiO and VO along with the strengthening of the Rb I and Cs I lines toward later L types. The K I lines at 7665 and 7699 Å have cores that slowly broaden through L4, where they virtually disappear, and wings that become increasingly more prominent for the latest L types. Note also that the CrH band at 8611 Å strengthens from late-M through L5 then weakens again thereafter. As shown in the full spectra of Figure kirfig3, the slope of the spectrum stays roughly constant for early-L dwarfs then increases noticeably for mid- and late-L dwarfs.

Spectral indices were then developed to measure these changes quantitatively. The definitions of these indices are given in [11] along with the complete recipe for classification[3]. Fig. 5 shows measured indices for all 25 L dwarf spectra and some late M dwarfs to illustrate the correlation between an individual index and the final, assigned spectral subtype.

5 Using the System to Type New Discoveries

A major test of any classification system comes with the typing of new objects discovered only after the classification scheme was introduced. We have now extended the classification recipe of [11] to 92 L dwarfs[4]. Measured spectral indices for all 92 L dwarfs are illustrated in Fig. 6. As marked in Fig. 6, only *two* of those 92 spectra have measured spectral indices that imply grossly conflicting spectral subclasses [13]. This, however, is a triumph of the classification scheme because a look at these spectra reveals obvious peculiarities – i.e., they look unlike any other L dwarfs currently known. In short, the classification scheme has proven to work well.

---

[3] The denominator of the TiO-b index should be 8455.0-8470.0 Å, not 8435.0-8470.0 Å listed in the paper.

[4] For spectra with lower signal-to-noise, the narrow indices that measure the strengths of the Rb I and Cs I lines are less accurate, so [13] provides a slightly augmented recipe for the classification of such spectra.

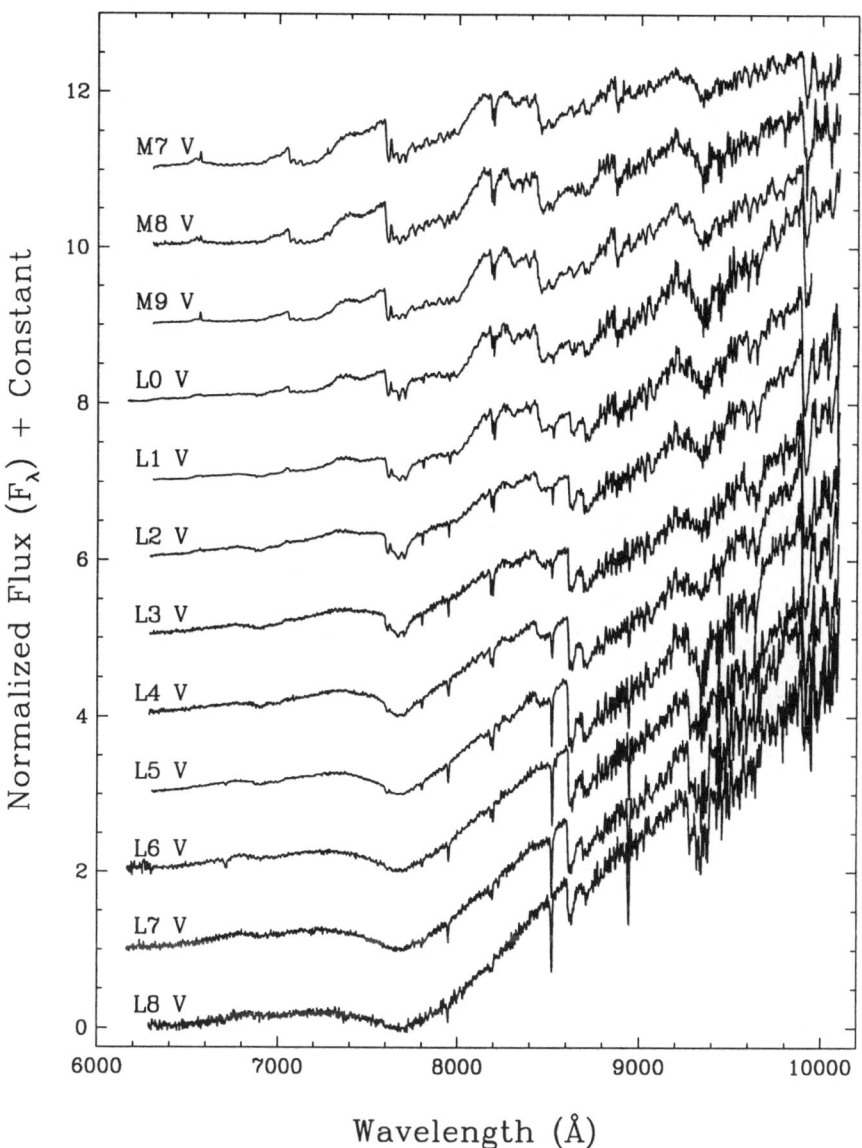

Fig. 3. The far red spectral sequence from M7 through L8 showing the slow "morphing" of spectra from one subtype to the next.

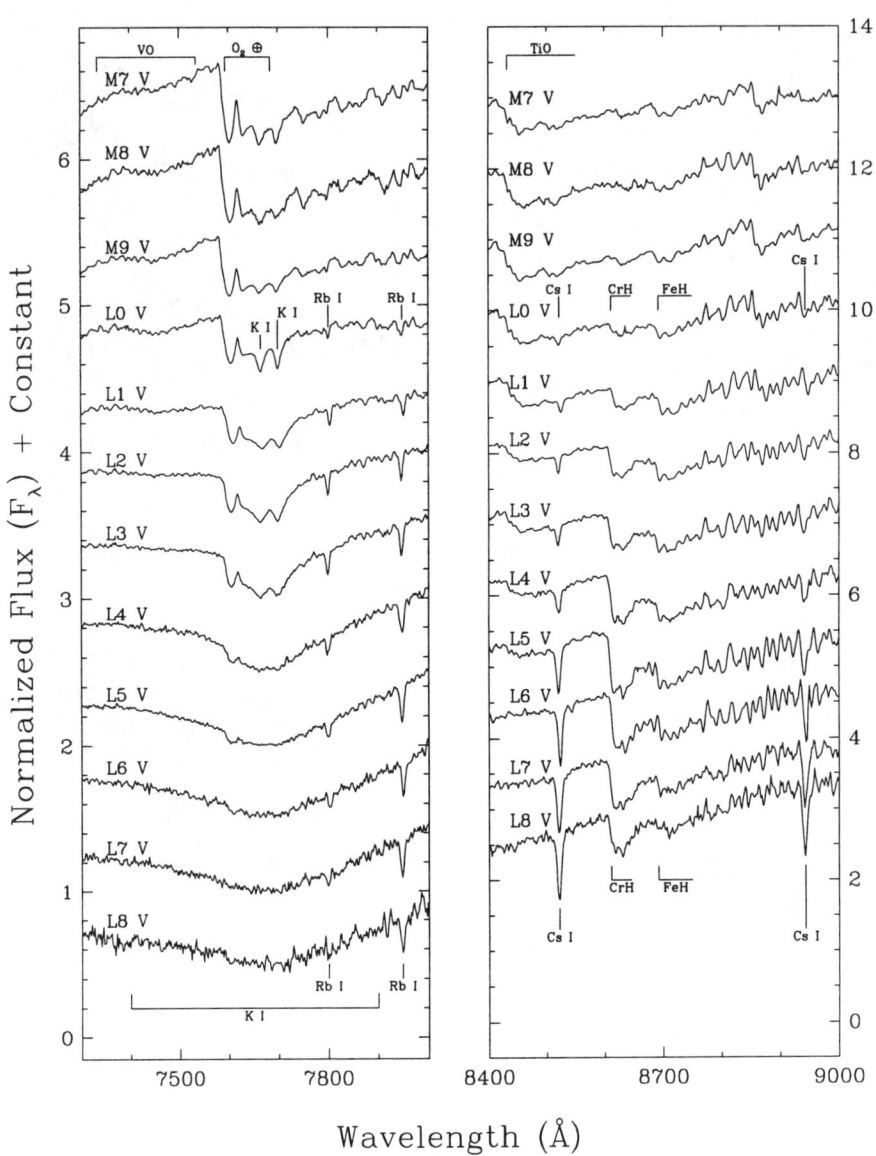

Fig. 4. Details of the spectra plotted in Fig. 3 showing changes in line and band strengths in the 7300-8000 Å interval (left) and the 8400-9000 Å interval (right). Prominent features are marked.

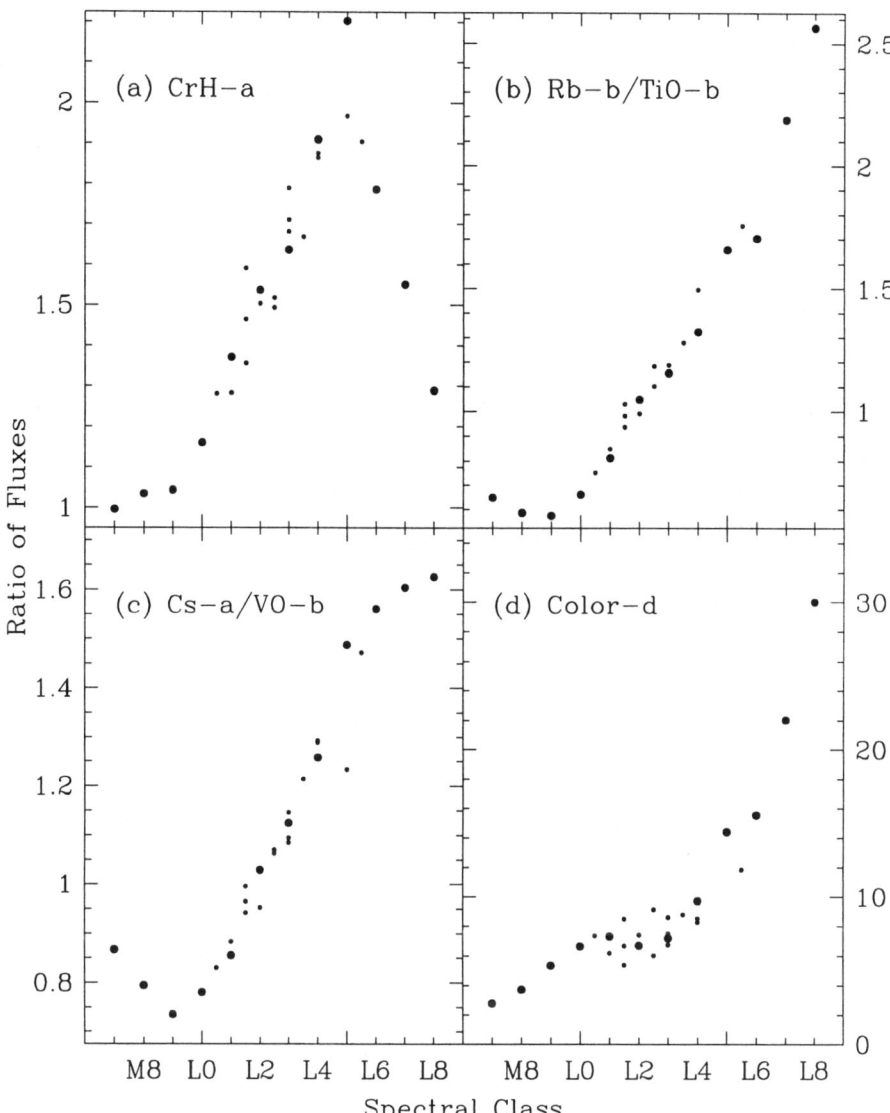

Fig. 5. Measured values for four classification indices plotted as a function of spectral type. Objects chosen as spectral standards are denoted by larger dots.

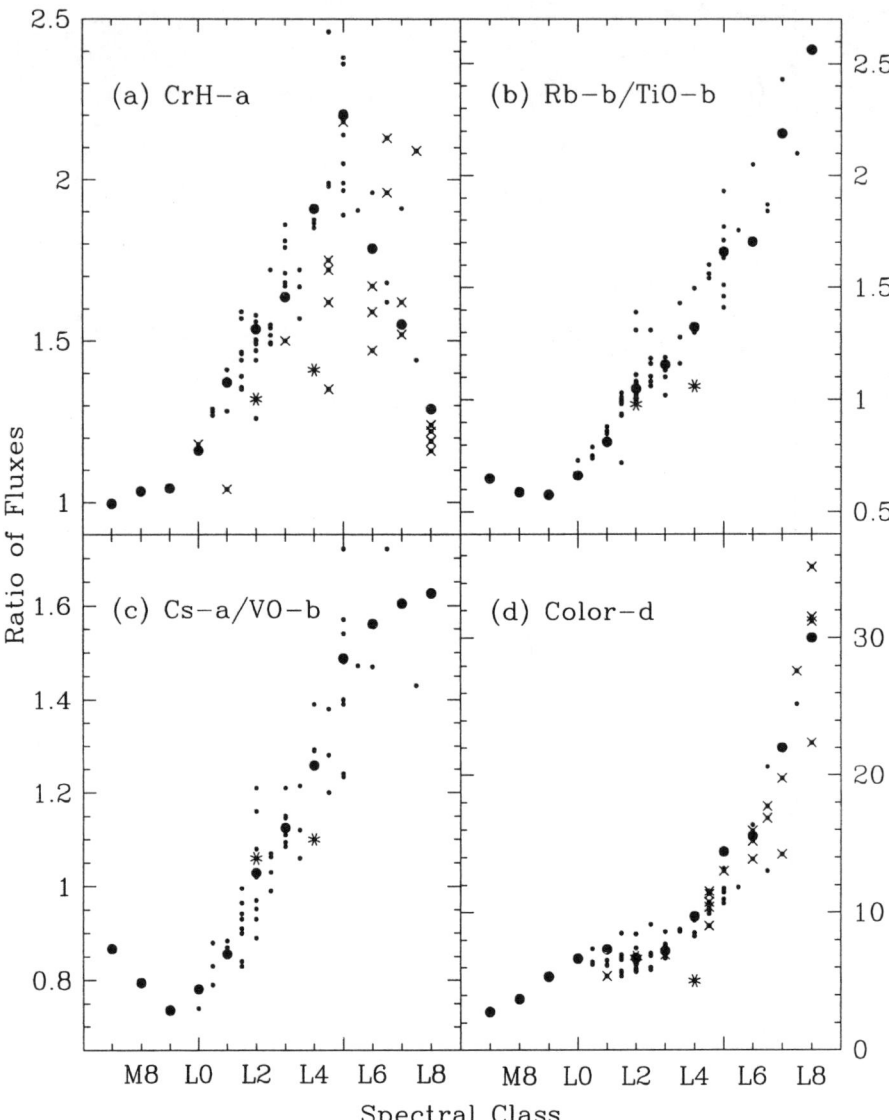

Fig. 6. Measured values for the four spectral indices of Fig. 5, but this time for the full set of 92 L dwarf spectra. Large dots again denote standards. Crosses in panels (**a**) and (**d**) denote spectra with lower signal-to-noise. The two starred symbols denote two peculiar spectra, both of which have fits to their K I lines that are inconsistent with the type implied by indices measured here. See [13] for details.

Concerns were raised that extending the L dwarf classification to L8 might leave too little room (just subtypes L8.5, L9, and L9.5) between the latest known L dwarfs and very early T dwarfs. Despite surveying vast amounts of the 2MASS sky in search of L dwarfs later than L8, we have yet to find any examples [13]. All later objects are clearly T dwarfs. Fig. 1b shows the near-infrared colour-colour diagram as it is known today with all 92 L dwarfs from 2MASS shown along with all known T dwarfs having three-band 2MASS detections. Early-L dwarfs are located near $J - K_s \approx 1.2$ in the plot, with late-L dwarfs extending to $J - K_s \approx 2.1$. Redder than this no later L dwarfs have been found despite extensive searching. We believe that at temperatures slightly cooler than L8, methane begins to develop in the H and K bandpasses, sending the colours toward the blue. Two early-T dwarfs (open stars) with weak methane features are seen near $(J - H, H - K_s) \approx (0.9, 0.2)$. For even cooler T dwarfs (also shown by open stars), deeper methane bands produce colours near $J - K_s \approx 0.0$.

If L dwarfs later than L8 should be discovered, we have room in the sequence to accomodate them. However, the evidence is growing that types between L8 and T0 may not even be used – physical calculations of [13,24] argue that the L8 dwarf Gl 584C [14] has a temperature near 1300K, close in teperature to the upper end of the T dwarf sequence [27].

6 Extending the Classification to Other Wavelength Regimes

The 6300-10000 Å spectral region is rich in spectral diagnostics, but the drawback is that L dwarfs are very faint at these wavelengths. Therefore, it would be desirable to establish classification also in the near-infrared (1-2.5 μm) where the flux is greatly enhanced relative to the optical. This is currently being pursued by a number of groups. Preliminary indicators show that ordering by spectral morphology independently gives the same sequence in the near-infrared as it does in the optical [18,28].

Extending L dwarf classifications into the near-infrared is desirable also for another very important reason: the defining feature of T dwarfs (the onset of methane at K-band) falls at those wavelengths. As reported elsewhere in this volume [4], it is shown that L dwarfs and T dwarfs are distinguishable from one another in both wavelength regimes. Hence, it has been established that an object will not be classified, for example, as an L dwarf in the far optical but as a T dwarf in the near-infrared.

7 Conclusions

The philosophy behind our classification scheme has been described. Although based originally on only 25 L dwarf spectra, we have shown that the scheme works well on a much large sample of 92 L dwarfs. Thus, our classification provides a robust framework against which to categorize newly-discovered objects. Now that the far red spectral classification has successfully bridged the span from M dwarfs through L dwarfs, the next step is to develop typing schemes for L and T dwarfs using spectral morphology at near-infrared wavelengths. Adam Burgasser picks up this part of the story in his contribution to these proceedings [4].

Appendix: An Alternative L Classification?

Alternative classifications have been provided for a small number of L dwarfs by [16,1]. These authors prefer not to use spectral morphology as the basis of their classification, but instead attempt to tie their classifications into a temperature scale based on fits to spectral synthesis models. Their classification is based on fits [1] to a single high-resolution line – Cs I at 8521 Å. (They dismiss the fits for the Rb I at 7948 Å, which gives temperature estimates systematically higher by 100-200K than those of Cs I.) In cases where only low-resolution data have been acquired, the classification relies on a single psuedocontinuum-slope [16] tied to the grid of types assigned by [1].

It has been stated in the literature [16] that the Basri/Martín types agree with those of Kirkpatrick for early L dwarfs. We do not agree. The philosophy behind their scheme and ours is totally different. As a case in point, consider the two spectra shown in Figure 16 of [16]. Both are typed as L1 under the Basri/Martín scheme. We have independent spectra of one of the objects (G 196-3B) and find a type for it on our system of L2. The other object, DENIS-P J1441-0945, would get a type around L0 on our system based on the spectrum published in [16]. Based on the spectral morphology of these two objects, we contend that they are quite different and should be placed into different – not the same – subtypes.

Other disadvantages of the Basri/Martín system are –

- The exact recipe for setting up the grid of types based on the model fits is unclear. Inexplicably, spectra which have equal temperature estimates based on the model fitting are in some cases placed into *different* subtypes. See Table 5 of [16] and Table 2 of [1].
- An object typed using the pseudocontinuum ratio of [16] could receive a far different classification if the model fits of [1] are used instead.
- The pseudocontinuum index of [16] is essentially single-valued in the region between L0 and L3, so classifications based on this index alone are at the mercy of random errors associated with noise. Note that our own Colour-d index in Figs. 5 and 6 has the same problem which is one of many reasons why we chose not to rely solely on that measurement.
- Readers should also be aware that types on the Basri/Martín system may change as model atmospheres are improved.

Concerning this last point, the authors refer to their types as "preliminary" [16], but the bottom line is that the types are not immutable. Models will, of course, continue to be improved into the forseeable future. As a result we would have ever-changing spectral types under the Basri/Martín system. It is easy to imagine the confusion this would cause if an L4 is reassigned to an L5 in the future and perhaps re-reassigned to another type even later. An astronomer speaking of an L5 spectrum would have to indicate clearly the year in the which the object was typed and/or the pedigree of the theoretical models used in the typing. It should be stated that the reason for classification is to aid scientists in getting to the eventual answer – an understanding of the physics which shapes the spectral energy distributions as we see them. Any system which tries to use physical underpinnings as the *basis* of the classification is simply misdirected in its goals.

References

1. G. Basri, S. Mohanty, F. Allard, P. H. Hauschildt, X. Delfosse, E. L. Martín, T. Forveille, B. Goldman: ApJ, **538**, 363 (2000)
2. E. E. Becklin, B. Zuckerman: Nature, **336**, 656 (1988)
3. M. S. Bessell, J. M. Brett: PASP, **100**, 1134 (1988)
4. A. J. Burgasser, J. D. Kirkpatrick, M. E. Brown: this volume (2001)
5. A. J. Cannon, & E. C. Pickering: Ann. Astron. Obs. Harvard Coll. **28(II)**, 131 (1901)
6. X. Delfosse, et al.: A&A, **327**, L25 (1997)
7. J. D. Kirkpatrick, T. J. Henry, D. W. McCarthy, Jr.: ApJS, **77**, 417 (1991)
8. J. D. Kirkpatrick, T. J. Henry, J. Liebert: ApJ, **406**, 701 (1993)
9. J. D. Kirkpatrick, C. A. Beichman, Skrutskie, M. F.: ApJ, **476**, 311 (1997)
10. J. D. Kirkpatrick: 'Spectroscopic Properties of Ultra-cool Dwarfs and Brown Dwarfs'. In *Brown Dwarfs and Extrasolar Planets, held at Puerto de la Cruz, Tenerife, Spain, March 17-21, 1997*, ed. by R. Rebolo, E. Martín, M. R. Zapatero Osorio (ASP, San Francisco 1998) pp. 405-415
11. J. D. Kirkpatrick, et al.: ApJ, **519**, 802 (1999)
12. J. D. Kirkpatrick, F. Allard, T. Bida, B. Zuckerman, E. E. Becklin, G. Chabrier, I. Baraffe: ApJ, **519**, 834 (1999)
13. J. D. Kirkpatrick, et al.: AJ, **120**, 447 (2000)
14. J. D. Kirkpatrick, J. E. Gizis, A. J. Burgasser, J. C. Wilson, C. C. Dahn, D. G. Monet, I. N. Reid, J. Liebert: this volume (2001)
15. S. K. Leggett: ApJS, **82**, 351 (1992)
16. E. L. Martín, X. Delfosse, G. Basri, B. Goldman, T. Forveille, M. R. Zapatero Osorio: AJ, **118**, 2466 (1999)
17. K. Matthews, T. Nakajima, S. R. Kulkarni, B. R. Oppenheimer: AJ, **112**, 1678 (1996)
18. I. S. McLean, et al.: ApJ, **533**, 45 (2000)
19. W. W. Morgan, P. C. Keenan: ARA&A, **11**, 29 (1973)
20. T. Nakajima, B. R. Oppenheimer, S. R. Kulkarni, G. A. Golimowski, K. Matthews, S. T. Durrance: Nature, **378**, 463 (1995)
21. J. B. Oke, et al.: PASP, **107**, 375 (1995)
22. B. R. Oppenheimer, S. R. Kulkarni, K. Matthews, T. Nakajima: Science, **270**, 1478 (1995)
23. E. C. Pickering: Ann. Astron. Obs. Harvard Coll. **27**, 1 (1890)
24. I. N. Reid, J. E. Gizis, J. D. Kirkpatrick, D. W. Koerner: AJ, **121**, 489 (2001)
25. M. T. Ruiz, S. K. Leggett, F. Allard: ApJ, **491**, L107 (1997)

26. C. Sagan, *Cosmos* (Random House: New York 1980), p. 93
27. D. Saumon, A. J. Burgasser, M. S. Marley: priv. comm. (2000)
28. J. C. Wilson, J. R. Houck, M. F. Skrutskie, J. E. Gizis, D. G. Monet, A. J. Burgasser, J. D. Kirkpatrick: BAAS, **32**, 678 (2000)
29. B. Zuckerman, E. E. Becklin: ApJ, **386**, 260 (1992)

Spectroscopy of Young Brown Dwarfs and Isolated Planetary Mass Objects

E.L. Martín

Insitute for Astronomy, University of Hawaii, 2680 Woodlawn Drive, Honolulu, HI 96822, USA

Abstract. This paper presents an overview of the spectroscopic properties of substellar objects found in young (age between 1 and 20 Myr) open clusters and stellar associations. An updated status of spectral classification and temperature calibrations is given. A reddening-independent spectral index is provided for L-type dwarfs. The spectral types that separate stars from brown dwarfs, and brown dwarfs from planetary mass bodies, are discussed. The behaviour of gravity sensitive features, which potentially can play a key role in an age scale for star forming regions and young open clusters, is described. Current one-dimensional schemes for spectral classification of ultracool dwarfs should be taken with caution. The observed emission lines in young brown dwarfs and isolated superplanets are discussed. A possible selection bias against finding accreting brown dwarfs and isolated superplanets (substellar equivalents of the classical T Tauri stars) is identified. Finally, we comment on the possibility of using deuterium burning to provide an age scale for young associations and clusters.

1 Introduction

The early phase of contraction of brown dwarfs and planetary mass bodies is the most favourable for detecting them because of their relatively high luminosity and temperature. For simplicity, we arbitrarily adopt the "nuclear" terminology which defines a brown dwarf as an object unable to stabilize on the hydrogen-burning main sequence, and a planetary mass body as an object unable to produce any fusion reactions in its interior (Saumon et al. [50]; Oppenheimer et al. [44]). The adopted definition of planetary mass object is independent of how the object forms. The mass is what really counts (and to a lesser extent chemical composition and rotation) because it determines the structure of the object as a function of time. The practical boundary between stars and brown dwarfs is at 0.075 $M_\odot$ (78 M_{Jup}) for solar composition (Baraffe et al. [2]) and the boundary between brown dwarfs and planetary mass objects is at 12 M_{Jup} (Saumon et al. [50]). For ages younger than 20 Myr, the substellar boundary is thought to be at about spectral type M6 (Luhman [29]; Martín et al. [34]). On the other hand, the spectral type of the brown dwarf-planetary mass boundary is very age-dependent as discussed in Section 2.

The first brown dwarf in a star-forming region was identified in the Rho Ophiuchi cloud by Rieke & Rieke [48]. Ironically, they did not realize it because the object has low reddening. Optical low-resolution spectroscopy of this object, namely Rho Oph J162349.8-242601, was presented by Luhman et al. [28]. They classified it as an M8.5 spectral type and found strong H$_\alpha$ emission. Martín et al. [34] definitely established the substellar mass of this object by reporting the detection of the LiI resonance feature. They also measured a photospheric radial velocity consistent with the CO cloud cores velocities in Rho Oph and an H$_\alpha$ emission equivalent width as strong as that measured by Luhman et al. Cushing et al. [14] obtained a low-resolution H and K spectrum and estimated a spectral type of M8 from the strength of the water-absorption bands (Q-index).

The activity in the field of young brown dwarfs has accelerated in the last year: Comerón et al. [13] have presented optical spectroscopy for brown dwarfs in the Chamaeleon I star formation region. They derived spectral types for the objects. Seven of them are in the range M7 to M8. They also presented near-infrared spectra for a subset of them. They did not find any apparent correlation between spectral subtype and the strength of the water bands. Low-resolution optical spectra for brown dwarfs as late as M8.5 in the the young open cluster IC 348 has been presented by Luhman [29]. Another young open cluster, Sigma Orionis, has been subject to deep searches for brown dwarfs and isolated planets. Béjar et al. [6] obtained low-resolution optical spectra for three objects that they classified between M7 and M8. Zapatero Osorio et al. [57] presented low-resolution optical and H and K spectra of four objects that they classified as L dwarfs. They argued that the masses of these objects are below the deuterium burning mass threshold. Ardila et al. [1] carried out a survey for low-mass members of the Upper Scorpius OB association. They obtained low-resolution optical spectra of 22 objects. Seven of them have spectral types between M6 and M7.

This paper is organized as follows. Section 2 gives an update status of spectral classification and temperature calibrations and Section 3 describes the behaviour of gravity sensitive features. Section 4 discusses the observed emission lines with emphasis on the differences between young brown dwarfs and other types of emission line low mass objects.

2 Spectral Types and Temperatures

Spectral types for young brown dwarfs are usually inferred by comparing their spectrum with field dwarfs and giants. A library of low-resolution optical spectra of ultracool dwarfs (spectral types from M8 to L6) is available on the web[1]. Martín et al. [35] have calibrated one pseudocontinuum index (PC3) versus spectral type[2] from M2.5 to L6. The index measures points that are 700 nm apart. Thus, reddening effects in star-forming regions may be significant. A reddening-independent index may be defined using a standard interstellar extinction law. Such an approach, analogous to the UBV Q index used to determine spectral types of early type stars (Johnson & Morgan [22]), has been adopted by Wilking

---

[1] http://www.ucm.es/info/Astrof/fgkmsl/mldwarfs.html.

[2] An independent spectroscopic classification scale for L dwarfs has been given by Kirkpatrick et al. [25]. It is based on higher resolution optical spectra than the Martín et al. [35] scale.

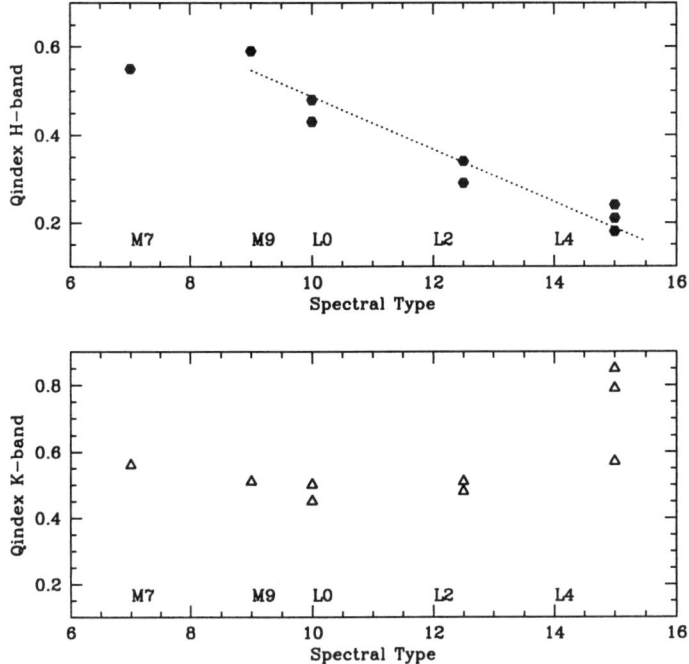

Fig. 1. Upper panel: Values of the Q index in the H-band defined in this paper versus optical spectral types in the Martín et al. 1999b scale. Lower panel: Values of the Q index in the K-band defined by Wilking et al. (1999) versus optical spectral types in the Martín et al. 1999b scale.

et al. [54] for K-band spectra, but their Q index saturates at about M8 and is not useful for L dwarfs.

For young substellar objects, which are frequently found in regions of high extinction, it is useful to define a reddening independent index for low-resolution near-infrared spectra that is sensitive to temperature for L dwarfs. Here, we define a Q index in the H-band, which provides a fairly good correlation with optical spectral types in the range M9 to L5 (Fig. 1). This index is defined as: $Q_{H-band} = ((F1/F2) \times (F3/F2))^\beta$ where F1, F2 and F3 are the average values of the relative flux densities computed in the three bands 1.51-1.56, 1.66-1.71 and 1.75-1.80 μm, respectively; $\beta=[A(1.535\ \mu m)-A(1.685\ \mu m)]/[A(1.775\ \mu m)-A(1.685\ \mu m)]$; and $A(\lambda)$ is the extinction at wavelength λ. Using the Draine [17] extinction law $A(\lambda) \sim \lambda^\alpha$ with $\alpha=-1.75$, we obtain $\beta=2.036$. Use of other extinction laws does not change the spectral types by more than 0.2 subclasses. We have derived a relationship between the Q index in the H-band and optical spectral type (in the Martín et al. [35]) using spectra obtained with three different instruments (CGS4 on UKIRT, Leggett et al. [27]; IRIS on the 3.9-m Anglo Australian telescope, Delfosse et al. [16]; and NIRC on Keck I). The standards used were LHS 3003 (M7), LHS 2065 (M9), DENIS-P J0909-0658 (L0), DENIS-P J1058-1548 (L2.5) and DENIS-P J0205-1159 (L5). The relationship is: $Subclass = 18.133 - 16.69 \times Q_{H-band}$ the rms is 0.61 subclasses. We note that a Q-index could also be defined for optical spectra, using a combination of the PC3 and PC6 indices defined by Martín et al. [35]

White et al. [55] have convincingly argued that the spectra of late M young brown dwarfs are intermediate between dwarf and giants and that an intermediate temperature calibration improves the comparison between observations and evolutionary models in the H-R diagram. The temperature scales of M dwarfs and giants are different. Luhman [29] derived an intermediate temperature scale. Béjar [7] has obtained a temperature calibration for members of the sigma Orionis cluster using spectral synthesis fitting of optical spectra (range 600–900 nm). Table 1 gives a comparison between the temperature calibrations for young brown dwarfs of Luhman [29] (L99) and Béjar [7] (B00), Pleiades brown dwarfs (Martín et al. 2001, in preparation - M01) and recent calibrations for field dwarfs (Jones et al. [23] - J96; Bessell et al. [9] - BS98; Basri et al. [5] - Ba00; Pavlenko et al. [45] - P00; Leggett et al. [27] - L01). Major discrepancies in temperatures are found at M6 (the B00 temperature is warmer by more than 300 K with respect to L99), M7 and M9 (the L01. scale is cooler than the others) and at L5 (the Ba00 is 550 K hotter than that of P00). For spectral subclasses between M9.5 and L4, the agreement among the various scales is better than 200 K. This could be a coincidence, or a hint that the spectra of those objects are simpler to interpret than that of the hotter M dwarfs. The differences among the temperature scales indicate that we are far from reaching a complete understanding of the atmospheres of ultracool dwarfs. Synthetic spectra are not able to fit all the observed features. An example is shown in Fig. 2, where we compare the spectrum of Teide 1 (a benchmark M8 brown dwarf in the Pleiades, Rebolo et al. [47]) with a theoretical spectrum provided by France Allard. The models are unable to reproduce the strength of the main molecular features (CO, TiO, steam). Similar problems may be seen in Ruiz et al. [49] and Leggett et al. [27]. Since the atomic features are formed against a strong background of molecular opacities, it is very difficult to measure chemical abundances.

The evolutionary models of Baraffe et al. [2] and Chabrier et al. [11] indicate that the substellar mass limit is at a temperature of 2932 K at 1 Myr, 3000 K at 5 Myr, 3006 K at 10 Myr and 2995 K at 20 Myr. Thus, according to all the spectral type versus temperature calibrations in Table 1, young dwarfs with spectral type M7 and later are substellar objects. The same models indicate that the planetary mass limit is at a temperature of 2265 K at 1 Myr, 1929 K at 5 Myr, 1682 K at 10 Myr and 1434 K at 20 Myr. Such temperatures straddle the whole L-dwarf domain. The recent results of Najita et al. [42] and Zapatero Osorio et al. [57] indicate that there is a numerous population of objects occupying the whole mass range of brown dwarfs in the IC 348 and sigma Orionis clusters, respectively. Spectroscopic observations of these objects will yield a spectral sequence of objects with common distance and metallicity and a narrow range of ages. In Table 2, we use the theoretical models (dust-free for temperatures

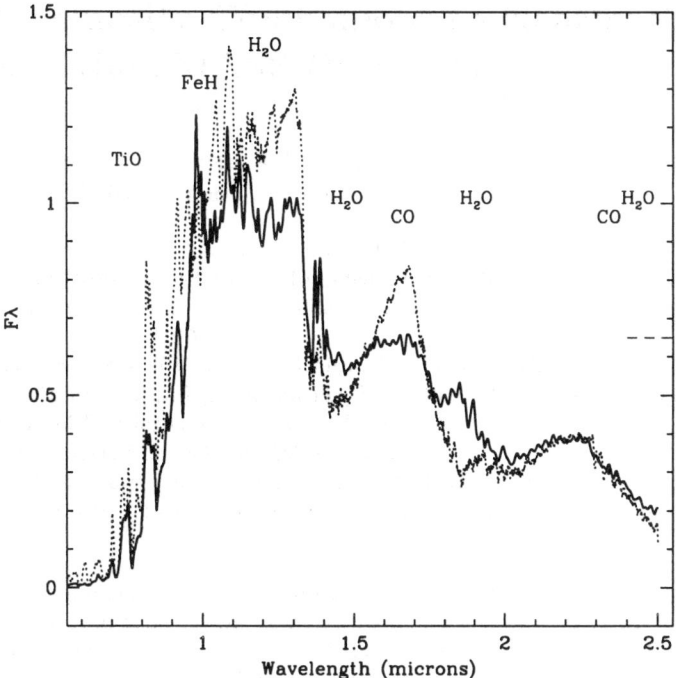

Fig. 2. Spectrum of Teide 1 (solid line) compared with a model spectrum (dotted line) provided by France Allard (solar metallicity, 2400 K, log g=4.0, dusty atmosphere, AMES water list).

above 2400 K, dusty for temperatures below, see Chabrier et al. [11] for more details) and temperature calibrations to illustrate the locations of the substellar and planetary boundaries in several interesting young open clusters and stellar associations. This is not intended to be an exhaustive list of young regions around the Sun but rather some examples of regions where the whole range of brown dwarf masses, and the associated L subtypes, may be observed. The ages that we have used are representative, but an age-range of a few Myr is likely to be present (Herbig [20]; Martín [31]).

The examples given in Table 2 indicate that it is feasible with current instrumentation on large telescopes to detect and obtain low-resolution spectra in the optical and near-infrared of brown dwarfs in the whole range of masses from the substellar to the planetary domain. Rich young associations and clusters such as Sigma Ori and Upper Sco provide the unique opportunity of studying a population of objects covering the whole range of L subclasses and the transition to the T (H) class[3] (Leggett et al. [26]). Young isolated planets with masses similar to Jupiter should be T (H) dwarfs and could be recognized by their peculiar

---

[3] The "T "class was defined by Kirkpatrick et al. [25] as that of objects similar to Gl 229 B, which have strong methane bands in the near-infrared spectrum (Oppen-

Table 1. Spectral types and effective temperatures for cool dwarfs (see text for explanation of column headings)

SpT	B00	L99	J96	BS98	Ba00	P00	L01	M01
M6	3300	2990	2900	2720	2800			2800
M7	2800	2890		2550			2250	2700
M8	2700	2720	2750	2520				2450
M9	2550	2550	2350	2400	2375		2100	2350
M9.5					2300	2200	2100	
L0					2200		2000	2150
L1	2000				2100			
L2					2000	2000	1900	
L3					1950		1900	
L4					1850		1900	
L5					1750	1200	1850	
L6					1700			

Table 2. Location of the substellar and planetary mass boundaries for young associations and clusters

Name	Dist. (pc)	Age (Myr)	Boundary	$\log(L/L_\odot)$	I (mag)	J (mag)	T_{eff} (K)	SpT
Taurus	140	1	Substellar	-1.39	14.1	11.9	2932	M6
Taurus	140	1	Planetary	-2.82	18.5	15.3	2265	L0
Sigma Ori	360	5	Substellar	-1.79	17.0	15.0	3000	M6
Sigma Ori	360	5	Planetary	-3.42	22.9	19.1	1929	L3
Upper Sco	140	10	Substellar	-2.05	15.5	13.6	3006	M6
Upper Sco	140	10	Planetary	-3.75	22.5	18.5	1682	L6

infrared colours. Comparison of these objects to the field T (H) dwarfs would yield important clues to the gravity dependence of methane and steam bands.

---

heimer et al. [44]). Martín et al. [35] suggested that a better choice of letter would be "H" because T dwarfs could lead to confusion with T Tauri stars and T associations.

3 Pitfalls of One-Dimensional Classification Schemes for Ultracool Dwarfs: Gravity Sensitive Features

Current schemes for spectroscopic classification of ultracool dwarfs are one dimensional (Kirkpatrick et al. [25]; Martín et al. 1999 [35]). They try to classify the objects only as function of effective temperature. However, gravity is known to change the spectral features. This is the reason why luminosity classes were introduced in the classical MK scheme (Morgan et al. [39]).

The gravity of substellar objects depends on the age and on the mass ($g \sim M^{-1/3} t^{2/3}$). For example, for a star-forming region where the typical age may be 1 Myr, the gravity of brown dwarfs range from log g=3.5 (0.075 $M_\odot$) to log g=3.8 (0.020 $M_\odot$). For a young open cluster of age 20 Myr, the gravities would vary between log g=4.4 and 4.2 for the same masses. For field objects, with typical ages of 5 Gyr, gravities range from log g=5.4 to 4.8 for the same masses. Thus, gravity may change by two orders of magnitude over the evolution of a brown dwarf. Substellar objects of similar temperature may have very different gravities. For a typical temperature of an L dwarf, i.e. 1600 K (Basri et al. [5]), gravity can vary from log g=4.0 (0.010 $M_\odot$ object at 10 Myr) to log g=5.4 (0.070 $M_\odot$ object at 5 Myr). Future spectroscopic schemes for ultracool dwarfs should be two-dimensional in order to account for differences in gravity for a given temperature. A calibration of gravity sensitive spectral features could provide a suitable age scale for young brown dwarfs and isolated planets.

In Table 3 a list of gravity sensitive spectral features is given (with comments on their dependence with gravity) and the range of spectral types for which such dependence has been observed. In general, the lines due to neutral alkali elements are weaker in young objects than those in field dwarfs. The TiO bands appear to be stronger for low gravity, but the hydrides (CaH, CrH, FeH) are weaker (Briceño et al. [10]; Martín et al. [35]). On the other hand, the VO bands at

Table 3. Gravity sensitive features in the optical spectrum

Feature	SpT	Comment
NaI589.00	L1-L2	Weaker for lower gravity
LiI670.8	M6-M8	Weaker for lower gravity
CaH697.5	M4-M7	Weaker for lower gravity
TiO705.4, 726.9	M4-M8	Stronger for lower gravity
KI766.49, 769.90	M4-L1	Weaker for lower gravity
NaI818.33, 819.48	M4-L1	Weaker for lower gravity
CsI852.13, 894.35	L1-L2	Weaker for lower gravity
FeH869.2, 989.6	L1	Weaker for lower gravity
CrH861.1, 996.9	L1	Weaker for lower gravity

785.1 nm and 853.8 nm have been reported to be rather insensitive to gravity in the spectral range M6 to L1 (Luhman [29]; Martín et al. [35]) . Thus, it seems that atomic lines should be used to calibrate gravity effects whereas VO bands should be used to calibrate the temperatures. However, the VO bands dissappear for late L dwarfs. As discussed in the previous section, steam bands may be used when VO is not available.

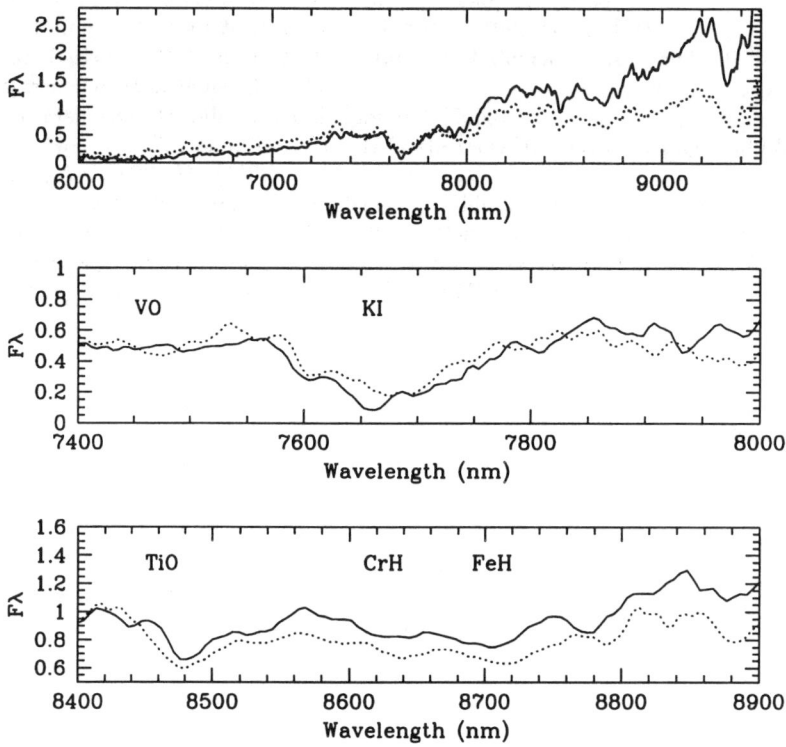

Fig. 3. Keck/LRIS spectra of SOri 47 (solid line) compared with Roque 25 (dotted line). In the upper two panels the spectra are shown normalized to 750 nm. In the lower panel the spectra are normalized to 840 nm. A boxcar smoothing of 5 pixels was applied.

Figures 3 and 4 illustrate the pitfalls of the one-dimensional-spectroscopic classification scheme for L dwarfs. They shows comparisons between optical and near-infrared spectra of SOri 47 (sigma Orionis member), Roque 25 (Pleiades member) and DENIS-P J0909-0658 (L0 dwarf in the field, Delfosse et al. [16]). All the objects were observed with the same instrumental configurations at the Keck observatory on Mauna Kea: the Low-Resolution Imaging Spectrograph (LRIS; Oke et al. [43]) and the Near-Infrared Camera (NIRC; Matthews & Soifer [37]).

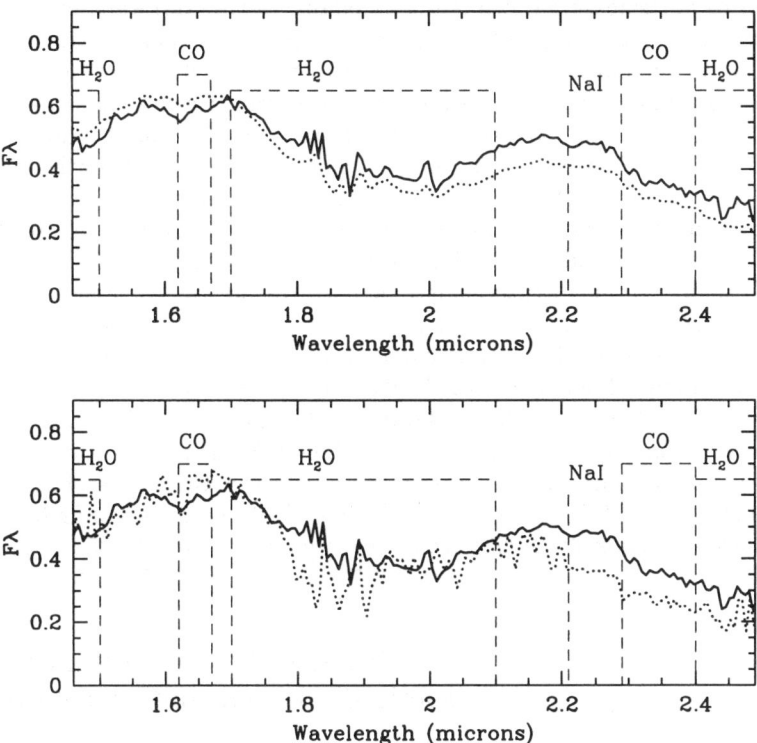

Fig. 4. Upper panel: Keck/NIRC spectra of SOri 47 (solid line) compared with DENIS-P J0909-0658 (dotted line). Lower panel: Keck/NIRC spectra of SOri 47 (solid line) compared with Roque 25 (dotted line).

Details of the LRIS observations are in Martín et al. [35]. The NIRC/HK grism gave a resolving power of $R=\Delta\lambda/\lambda=73$. Subtraction of telluric lines was performed using observations of an early G-type star observed at similar airmass as the program objects. The true continuum shapes of the objects were restored by multiplying by a Plank function with $T_{eff}=6000$ K. We compared our spectrum of DENIS-P J0909-0658 with the spectrum published by Delfosse et al. [16] and did not find any difference above the noise.

The comparison between these objects gives the following results: (1) SOri 47 has a steeper pseudocontinuum slope in the optical and broader KI resonance doublet than Roque 25. Thus, it has been assigned a spectral type of L1.5 (Zapatero Osorio et al. [56]), while Roque 25 was classified as L0 (Martín et al. [33]). Note, however, that the relative depth of the CrH and FeH at 861.1 nm and 869.2 nm with respect to TiO at 844 nm is stronger in Roque 25 than in SOri 47, implying that Roque 25 is cooler than SOri 47 in the Kirkpatrick et al. [25] scheme. Thus, the pseudocontinuum slope and KI feature give a relative

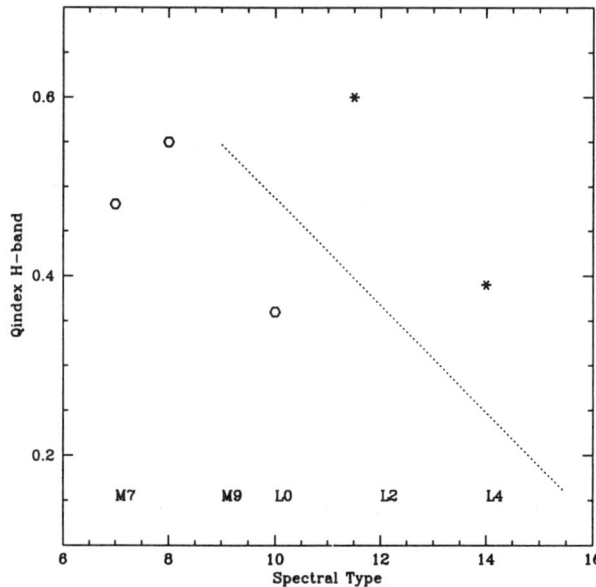

Fig. 5. Values of the Q index in the H-band for Sigma Orionis objects (six pointed stars, SOri 47 and 60, spectral types from Zapatero Osorio et al. [57] and Pleiades objects (Roque 15, Teide 1 and Roque 25; spectral types from Martín et al. [36]). The Q index - spectral subclass relationship for field dwarfs is plotted with a dotted line.

classification between these two objects that is contrary to the comparison between hydride and oxyde molecular features. (2) The CO bands at 1.619, 1.662 and 2.29 μm are stronger in SOri 47 than in DENIS-P J0909-0658. CO bands are known to be stronger in M-giants than in M-dwarfs (Greene & Meyer [19]), so this may be related to the low gravity of SOri 47. (3) The steam bands appear weaker in SOri 47 than in DENIS-P J0909-0658 and Roque 25. This could also be a gravity effect because the optical spectral type of SOri 47 is about 1 subclass later than the other two objects and steam bands are expected to become stronger for cooler temperatures, not weaker as observed.

The possible dependence of the steam bands with gravity implies that the Q index in the H-band is likely to be gravity sensitive. In Fig. 5 we plot measurements of the H-band Q index in Pleiades and Sigma Orionis objects. The Sigma Orionis objects tend to have weaker values than the field and Pleiades objects. More observations are required to calibrate this effect. The combination of spectroscopic features in the optical and near-infrared spectra of substellar objects in associations, clusters and the field, may lead to a two-dimensional classification scheme for late-M and L dwarfs.

4 Emission Lines: The Accreting Brown Dwarf Paradox

Gizis et al. [18] (also this volume) have presented a review of the activity in field ultracool dwarfs. They find that strong H_α emission does not imply youth. Young brown dwarfs and free-floating planets could, however, have an additional source of energy. Like their more massive analogs, the classical T Tauri stars, they could have accretion disks. One possible sign of disk accretion could be H_α emission stronger than that from field dwarfs. In Fig. 6 we compare the equivalent widths of young brown dwarfs with the upper envelope of the equivalent widths of field dwarfs (excluding flares and the unusual emission-line object PC 0025+0447, Schneider et al. [51]). The measurements for the field objects have been taken from the literature. We note that about half of the young brown dwarfs in IC 348, Sigma Orionis and Chamaeleon I have H_α equivalent widths stronger than the field dwarfs. This could be a sign of accretion or it could be due to enhanced chromospheric-heating. The latter explanation seems more likely because none of these dwarfs has H_α emission stronger than the field object PC 0025+0447 for which no accretion is required to explain the emission lines (Mould et al. [40]; Martín et al. [34]). Moreover, none of the young brown dwarfs has been reported to exibit [OI] and [SII] emission lines which are a hallmark of accretion processes in T Tauri stars (Muzerolle et al. [41]).

Since brown dwarfs and isolated planetary mass objects are thought to form as stars do, from the fragmentation and collapse of molecular clouds, we expect that they will have accretion disks during their early evolution. In fact, the disks may live longer around substellar objects because the radiation pressure and ultraviolet flux is much lower than in stars. However, the known young brown dwarfs and isolated superplanets do not present any clear signs of disk accretion (Béjar et al. [6], Luhman 1999 [29], Zapatero Osorio et al. [57]). There are no substellar equivalents to the classical T Tauri stars (CTTSs). This is a paradox because accreting brown dwarfs are expected to exist in star-forming regions; and they should outnumber the CTTSs if they accrete for a longer time and the IMF is similar to that of the Pleiades or sigma Ori clusters (Béjar et al. [8]; Martín et al. [36]). Two solutions appear possible: (1) Brown dwarfs do not form as stars do, so they do not accrete mass from disks, or the accretion is much faster than in stars (less than 10 Myr); (2) There is an observational bias against detecting accreting brown dwarfs. The second possibility needs to be explored because the effects of disk accretion have been shown to change the position of CTTSs in the H-R diagram (Kenyon & Hartmann [24]). The presence of a hot boundary layer (a narrow region where about half of the accretion luminosity is dissipated when disk material falls onto the central object and loses angular momentum) could shift blueward the optical colours of brown dwarfs. Accreting substellar objects would then appear to have blue colours and would be lost in the cloud of background stars that dominate the colour-magnitude diagrams of star-forming regions and young open clusters. We suggest that these objects could be identified because they are likely to be irregular, large-amplitude variables (an example could be the deeply embedded low-luminosity outbursting source found by Hodapp et al. [21] in Serpens), as their more massive counterparts, the

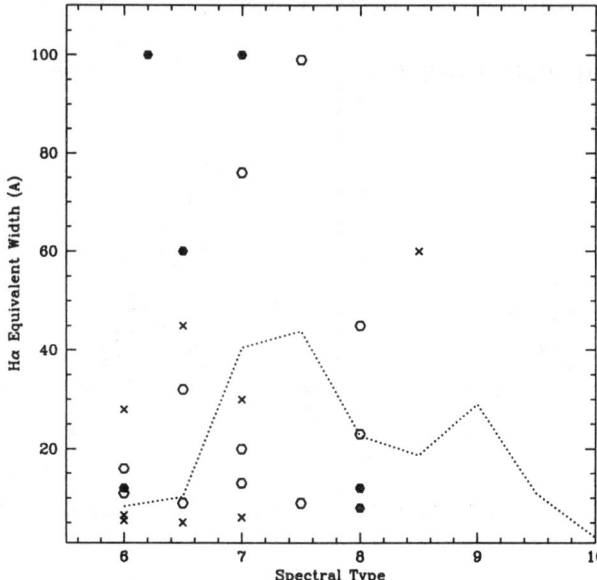

Fig. 6. The H_α equivalent widths of young brown dwarfs (four pointed stars, Sigma Orionis, Béjar et al. [6]; filled hexagons, IC 348, [29]; open hexagons, Chamaeleon I, Comerón et al. [13]) compared with the upper envelope of H_α equivalent widths of field dwarfs. One of the IC 348, an M6.2 dwarf, is outside the plot because it has H_α equivalent width of 250 A.

CTTS. Very deep infrared imaging surveys could also find them in regions of high obscuration, such as the cores of molecular clouds (an example could be some of the sources identified by Wilking et al. [54] in rho Oph). Finally, accreting brown dwarfs could have very strong H_α emission and might be detected with deep H_α imaging of molecular clouds.

5 Deuterium Dating

The term "Deuterium test", in analogy to the lithium test for brown dwarfs (Magazzù et al. [30]), should be restricted to using deuterium as an observable to discriminate between brown dwarfs and planets. In practice, this test can only be applied for ages older than about 100 Myr because that is the timescale for deuterium depletion in the lowest-mass brown dwarfs (Saumon et al. [50]).

In this paper, we are dealing with young substellar objects; whereas the use of deuterium is to serve as an age indicator rather than as a test to distinguish between brown dwarfs and planets. Spectroscopic observations of very low-mass stars and brown dwarfs have provided new ages for three open clusters using the sharp boundary between lithium preservation and lithium depletion in fully convective objects. The first cluster to which this method was applied is the

Pleiades. The canonical age is 75 Myr from fitting the upper main-sequence turnoff with standard isochrones. Overshooting, however, could bring the age calculated from the high-mass stars to 150 Myr (Mazzei & Pigatto [38]). The lithium age is about 120 Myr (Martín et al. [32]; Stauffer et al. [52]) consistent with mild overshooting. In αPersei the canonical age was 50 Myr, and the lithium age is about 90 Myr (Basri & Martín [4]; Stauffer et al. [53]). In IC 2391, the canonical age is 35 Myr and the lithium age is about 50 Myr (Barrado y Navascués et al. [3]). The lithium observations also indicate that the age spread among the members of those open clusters is small (less than 10 Myr). For clusters younger than about 5 Myr, lithium becomes inefficient as an age indicator because neither stars nor brown dwarfs have had time to burn it (D'Antona & Mazzitelli [15]). At these early ages, the destruction of deuterium is efficient in stars and brown dwarfs. Recent models by Chabrier et al.[12] show that deuterium is depleted by a factor of 100 in brown dwarfs with masses in the range 0.075 and 0.04 $M_\odot$ at nearly constant effective temperature and ages between 1 and 5 Myr. Thus, differential comparison between spectra of brown dwarfs of the same spectral type in regions like Chamaeleon, IC 348, Orion or Taurus could provide the means to detect variations in deuterium abundance. Furthermore, Chabrier et al. [12] and Pavlenko [46] predict that observable signatures (at the few percent level) are produced by HDO lines ontop of steam bands in the near-infrared spectrum. With the increasing number of brown dwarfs being discovered in young clusters and associations and the advent of high-resolution spectrographs on large-aperture telescopes (such as NIRSPEC on Keck, SPECS on IRTF and IRCS on Subaru), it seems that the application of the deuterium-dating method may bear fruit in the near future.

For associations and clusters with ages between 5 and 20 Myr, such as the Scorpius-Centaurus OB association, both deuterium and lithium are depleted in brown dwarfs although at different masses. Observations of both elements can provide complementary age indicators and test the accuracy of the model predictions.

6 Conclusions

We have presented an overview of the spectroscopic properties of young brown dwarfs and isolated superplanets (ages between 1 and 20 Myr, masses between 78 and 12 Jupiters) and we reach the following conclusions:

- Reddening-independent spectral indices that provide a smooth relationship between spectral type and effective temperature may be derived from low-resolution near-infrared spectra of M and L dwarfs. The Q index in the K-band works well for M dwarfs but saturates for L dwarfs. We have defined a Q index in the H-band index that is sensitive to spectral subclass for field L dwarfs.
- The discrepancies between different temperature scales are large (>300 K) for late M dwarfs, but become smaller (<200 K) for L dwarfs, except at

the very cool end (L5). Current theoretical model atmospheres cannot fit the observed molecular features in young brown dwarfs. Despite of the problems, all temperature calibrations agree that young dwarfs with spectral type M7 or later should have substellar masses. The planetary mass boundary is located between L0 and L6 for ages from 1 to 10 Myr, respectively.

- It is well-documented that lines from neutral alkalis (Cs, Na, K, Li) weaken with decreasing gravity. Molecular bands of steam seem to also weaken making the H-band index gravity sensitive but the observations are still scarce. CO and TiO bands strengthen with decreasing gravity; whereas hydrides (CaH, CrH, FeH) weaken. VO bands appear to be rather insensitive to gravity. All these features need to be carefully calibrated with high-quality spectra. They can provide a spectroscopic age-scale of young associations and clusters. Future classification schemes of ultracool dwarfs (late-M, L and T (H)) should use a variety of spectral features to provide a two-dimensional grid that would account for both gravity and temperature variations accross the cool edge of the HR diagram. The current one-dimensional schemes for spectroscopic classification of ultracool dwarfs should be taken with a grain of salt.

- About half of the brown dwarfs in Chamaeleon I, IC 348 and Sigma Orionis have H_α equivalent widths stronger than field dwarfs of similar spectral type (M6–M9). However, none of them have H_α emission stronger than the field object PC 0025+0447 for which no accretion is required to explain the emission lines. There are no clear substellar counterparts to the CTTSs. We argue, however, that they should exist and could be numerous. The problem may be that they could have bluer colours than non-accreting brown dwarfs due to the presence of a hot boundary layer and they may have escaped identification in photometric surveys.

- Observations of deuterium depletion in brown dwarfs can provide a new age scale for young associations and clusters. This method of deuterium dating is sensitive for ages less than about 5 Myr, for which the lithium dating is not efficient. Between ages 5 and 20 Myr both the deuterium and lithium dating can be applied for different masses, providing complementary age indicators.

Acknowledgements

I thank France Allard, Isabelle Baraffe and Victor Béjar for private communications of unpublished results. Mike Cushing helped in the definition of the Q index. I acknowledge the multiple contributions of Rafael Rebolo and Maria Rosa Zapatero Osorio during many years of collaborative work on substellar objects in the Pleiades and Sigma Orionis. I am grateful to the organizers for their invitation to deliver a talk and for allowing me to contribute this paper even though I could not attend the meeting because of an unexpected problem with my work visa in the US.

References

1. D. Ardila, E. L. Martín, G. Basri: AJ, **120**, 479 (2000)
2. I. Baraffe, G. Chabrier, F. Allard, P. H. Hauschildt: A&A, **337**, 403 (1998)
3. D. Barrado y Navascués, J. R. Stauffer, B. Patten: ApJ, **522**, L53 (1999)
4. G. Basri, E. L. Martín: ApJ, **510**, 266 (1999)
5. G. Basri, S. Mohanty, F. Allard, P. H. Hauschildt, X. Delfosse, E. L. Martín, T. Forveille, B. Goldman: ApJ, **538**, 363 (2000)
6. V. J. S. Béjar, M. R. Zapatero Osorio, R. Rebolo: ApJ, **521**, 671 (1999)
7. V. J. S. Béjar: PhD Thesis, University of La Laguna, Spain (2000)
8. V. J. S. Béjar et al.: ApJ, submitted (2000)
9. M. S. Bessell, F. Castelli, B. Plez: A&A, **333**, 231 (1998)
10. C. Briceño, L. Hartmann, J. Stauffer, E. L. Martín: AJ, **115**, 2074 (1998)
11. G. Chabrier, I. Baraffe, F. Allard, P. Hauschildt: ApJ, **542**, 464 (2000a)
12. G. Chabrier, I. Baraffe, F. Allard, P. Hauschildt: ApJ, **542**, L119 (2000b)
13. F. Comerón, R. Neuhauser, A. A. Kaas: A&A, **359**, 269 (2000)
14. M. Cushing, A. T. Tokunaga, N. Kobayashi: AJ, **119**, 319 (2000)
15. F. D'Antona, I. Mazzitelli: ApJS, **90**, 467 (1994)
16. X. Delfosse, C. G. Tinney, T. Forveille, N. Epchtein, J. Borsenberger, P. Fouqué, S. Kimeswenger, D. Tiphène: A&AS, **135**, 41 (1999)
17. D. T. Draine: Proc 22nd Eslab Symposium, ESA SP-290, p. 93 (1989)
18. J. E. Gizis, D. G. Monet, I. N. Reid, J. D. Kirkpatrick, J. Liebert, R. J. Williams: AJ, **120**, 1085 (2000)
19. T. P. Greene, M. R. Meyer: ApJ, **450**, 233 (1995)
20. G. H. Herbig: ApJ, **497**, 736 (1998)
21. K. W. Hodapp, J. L. Hora, J. T. Rayner, A. J. Pickles, E. F. Ladd: ApJ, **468**, 861 (1996)
22. H. R. Johnson, W. W. Morgan: ApJ, **117**, 313 (1953)
23. H. R. A. Jones, A. J. Longmore, F. Allard, P. H. Hauschildt: MNRAS, **280**, 77 (1996)
24. S. J. Kenyon, L. W. Hartmann: ApJ, **349**, 197 (1990)
25. J. D. Kirkpatrick et al.: ApJ, **519**, 802 (1999)
26. S. K. Leggett et al.: ApJ, **536**, L35 (2000)
27. S. K. Leggett, F. Allard, T. R. Geballe, P. H. Hauschildt, A. Schweitzer: ApJ, in press (2001)
28. K. L. Luhman, J. Liebert, G. H. Rieke: ApJ, **489**, L165 (1997)
29. K. L. Luhman: ApJ, **525**, 466 (1999)
30. A. Magazzù, E. L. Martín, R. Rebolo: ApJ, **404**, L17 (1993)
31. E. L. Martín: AJ, **115**, 351 (1998)
32. E. L. Martín, G. Basri, J. Gallegos, R. Rebolo, M. R. Zapatero Osorio, V. J. S. Béjar: ApJ, **499**, L61 (1998a)
33. E. L. Martín, G. Basri, M. R. Zapatero Osorio, R. Rebolo, R. García López: ApJ, **507**, L41 (1998b)
34. E. L. Martín, G. Basri, M. R. Zapatero Osorio: AJ, **118**, 1005 (1999a)
35. E. L. Martín, X. Delfosse, G. Basri, B. Goldman, T. Forveille, M. R. Zapatero Osorio: AJ, **118**, 2466 (1999b)
36. E. L. Martín, W. Brandner, J. Bouvier, K. L. Luhman, J. Stauffer, G. Basri, M. R. Zapatero Osorio, D. Barrado y Navascués: ApJ, **543**, 200 (2001)
37. K. Matthews, B. T. Soifer: in Infrared Astronomy with Arrays, ed. I. Mc Lean (Dordrecht: Kluwer), 239 (1994)

38. P. Mazzei, L. Pigatto: A&A, 193, 148 (1988)
39. W. W. Morgan, P. C. Keenan, & E. Kellman: *An Atlas of Stellar Spectra, with an Outline of Spectral Classification*, Chicago, Univ. Chicago Press (1943)
40. J. Mould, J. Cohen, J. B. Oke, N. Reid: AJ, **107**, 2222 (1994)
41. J. Muzerolle, L. Hartman, N. Calvet: AJ, **116**, 455 (1998)
42. J. R. Najita, G. P. Tiede, J. S. Carr: ApJ, **541**, 977 (2000)
43. J. B. Oke et al.: PASP, **107**, 375 (1995)
44. B. R. Oppenheimer, S. R. Kulkarni, K. Matthews, T. Nakajima: Science, **270**, 1478 (1995)
45. Y. V. Pavlenko, M. R. Zapatero Osorio, R. Rebolo: A&A, **355**, 245 (2000)
46. Y. V. Pavlenko: this volume (2001)
47. R. Rebolo, M. R. Zapatero Osorio, E. L. Martín: Nature, **377**, 129 (1995)
48. G. H. Rieke, M. J. Rieke: ApJ, **362**, L21 (1990)
49. M. T. Ruiz, S. K. Leggett, F. Allard: ApJ, **491**, L107 (1997)
50. D. Saumon, W. B. Hubbard, A. Burrows, T. Guillot, J. I. Lunine, G. Chabrier: ApJ, **460**, 993 (1996)
51. D. P. Schneider, J. L. Greenstein, M. Schmidt, J. E. Gunn: AJ, **102**, 1180 (1991)
52. J. R. Stauffer, J. L., G. Schultz, J. D. Kirkpatrick: ApJ, **499**, L199 (1998)
53. J. R. Stauffer et al.: ApJ, **527**, 219 (1999)
54. B. A. Wilking, T. P. Greene, M. R. Meyer: AJ, **117**, 469 (1999)
55. R. J. White, A. M. Ghez, I. N. Reid, G. Schultz: ApJ, **520**, 811 (1999)
56. M. R. Zapatero Osorio, V. J. S. Béjar, R. Rebolo, E. L. Martín, G. Basri: ApJ, **524**, L115 (1999)
57. M. R. Zapatero Osorio, V. J. S. Béjar, E. L. Martín, R. Rebolo, D. Barrado y Navascués, C. A. L. Bailer Jones, R. Mundt: Science, **290**, 103 (2000)

The Classification of T Dwarfs

A.J. Burgasser[1], J.D. Kirkpatrick[2], and M.E. Brown[1]

[1] California Institute of Technology, MSC 103-33, Pasadena CA 91125, USA
[2] Infrared Processing and Analysis Center, MSC 100-22, Pasadena CA 91125, USA

"...one is forced to wonder where it will lead to, if everyone who works on stellar spectra also introduces a new classification..."
– Nils Duner (1899)

Abstract. We discuss methods for classifying T dwarfs based on spectral morphological features and indices. T dwarfs are brown dwarfs which exhibit methane absorption bands at 1.6 and 2.2 μm. Spectra at red optical (6300–10100 Å) and near-infrared (1–2.5 μm) wavelengths are presented, and differences between objects are noted and discussed. Spectral indices useful for classification schemes are presented. We conclude that near-infrared spectral classification is generally preferable for these cool objects, with data sufficient to resolve the 1.17 and 1.25 μm K I doublets lines being most valuable. Spectral features sensitive to gravity are discussed, with the strength of the K-band peak used as an example. Such features may be used to derive a two-dimensional scheme based on temperature and mass, in analogy to the MK temperature and luminosity classes.

1 Introduction

In the last decade, we have witnessed a flood of low-mass star and brown dwarf detections, due to improvements in infrared array technologies and the advent of large-scale, near-infrared sky surveys using these detectors. In 1995, the first brown dwarfs were discovered after decades of failed searches – the companion object Gl 229B [31] and the Pleiades brown dwarf Teide 1 [36]. Since then, nearly one hundred brown dwarfs have been detected as companions to nearby stars and in the field, most made by new sky surveys such as the Two Micron All Sky Survey (2MASS [40]), the Sloan Digital Sky Survey (SDSS [43]), and the Deep Near-Infrared Survey (DENIS [12]). Others have been identified photometrically in various cluster surveys (see contribution by E. Martin, these proceedings). These surveys have finally brought brown dwarfs out of the realm of the theoretical and into that of the observational.

Two new spectral classes have also been defined as a result of these low-mass discoveries. The *L spectral class*, comprised of objects cooler than M9.5V, is defined by [19] and is discussed in the contribution by J.D. Kirkpatrick in these proceedings. The *T spectral class*, defined by the presence of CH_4 absorption at 2.2 μm [19], is discussed here. The prototype for this class, Gl 229B, shows CH_4 and H_2O absorption bands throughout the near-infrared [34,35], exhibiting a spectral morphology similar to those of the giant planets [14]. Gl 229B remained the solitary T dwarf until mid-1999, when 2MASS, SDSS, and the New Technology Telescope Deep Field uncovered field analogs. To date, at least 23 T dwarfs have been identified [3,4,6,11,42,24], a number sufficient to begin consideration of spectral classification.

1.1 Spectral Classification of Cool Dwarfs

The classification of late-type dwarfs (which we define here as objects later than spectral type K7V) has been done for over one hundred years using a variety of features in photographic, red optical, and near-infrared spectra. Figure 1 diagrams the wavelength and spectral type ranges of a few major classification schemes related to the widely accepted MK process. The classification of M dwarfs began with the conversion of Secchi's type III stars [39] to class M by W. Fleming in the Henry Draper catalog [32]. This class was further divided by A.J. Cannon into subtypes Ma, Mb, Mc, and Md, based on their spectral morphologies in the photographic regime [10]. Work by several authors using different instrumental configurations extended M dwarf classification to decimal systems up to type M6 [16,22,23,30,17], based primarily on the strength of TiO bands as temperature discriminants. Classification up to type M9V was done by P.C. Boeshaar in the photographic [1,2] and J.D. Kirkpatrick in the red optical [21,20], based on the appearance of TiO, VO, and CaOH bands. The latter scheme also makes use of spectral slope from 6300–9000 Å as an additional temperature discriminant. Following the discovery of numerous objects later than type M9.5V, J.D. Kirkpatrick defined and classified the L spectral class in the red optical based on the appearance of weakening oxides and strengthening metal hydride bands and alkali lines [19]. Classification of the L class in the near-infrared is currently under investigation by various authors. The MK(K) system [30,17] makes use of additional spectral features (e.g., CaOH, Na D lines) as luminosity discriminants in order to create a two-dimensional scheme. Luminosity class distinctions have not been made for L dwarfs, as the coolest giant stars retain strong TiO and VO bands [19].

The schemes for T dwarf classification shown as dashed lines in Fig. 1 are the focus of this article, and are discussed more fully in [7]. In §2 we address red optical classification, based on a spectral sample of bright T dwarfs identified by 2MASS and SDSS. In §3 we discuss near-infrared classification of a larger sample, and discuss spectral features important for classification. In §4 we describe spectral features in the near-infrared which may depend on specific gravity, and could therefore append a second dimension to a temperature scheme, in analogy to the MK luminosity classes. We summarize our discussion in §5.

2 Red Optical Classification

L dwarfs are classified by [19] using 6300–10100 Å red optical spectra at 9 Å resolution, obtained with the Low Resolution Imaging Spectrograph (LRIS [33]) mounted on the Keck 10m Telescope. As shown in Fig. 6 of [19], this spectral

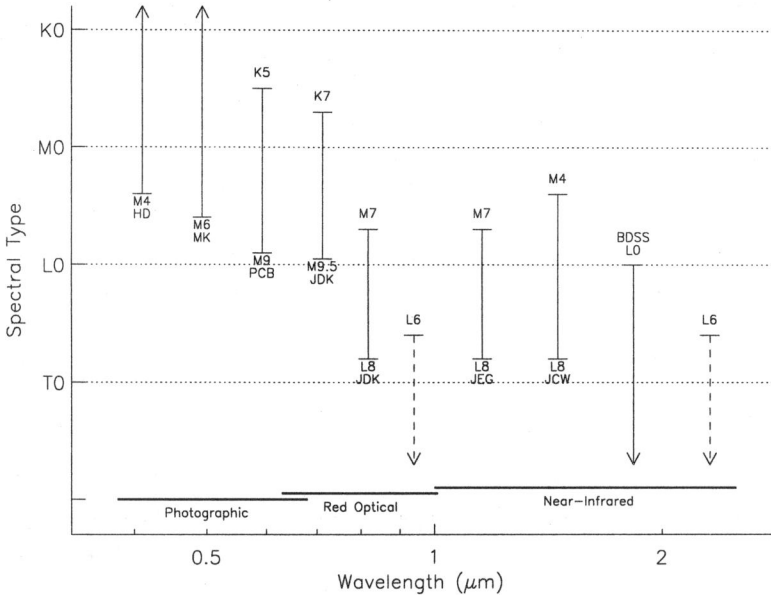

Fig. 1. Spectral typing schemes for M, L and T dwarfs over the past century. The HD scheme is that of the Henry Draper catalog [32,10,16]; the MK (Yerkes) scheme is described in [22,23,30,17]; PCB are the P.C. Boeshaar schemes of [1,2]; JDK are the M and L dwarf schemes of J.D. Kirkpatrick [21,20,19]; near-infrared schemes for L dwarfs are for J.E. Gizis (priv. comm.), J.C. Wilson (priv. comm.), and the NIR-SPEC Brown Dwarf Spectroscopic Survey (I. McLean; priv. comm.); the dashed lines represent spectral schemes discussed in the text, also described in [7].

range samples most of the prominent molecular and atomic features in late-M and L dwarf spectra, and a "recipe" for spectral classification from M7V to L8V is presented based on measurements of spectral indices on selected standards. In order to make a direct comparison with this L dwarf scheme, we have obtained optical spectra for nine T dwarfs utilizing the same instrumental setup as [19]. Data acquisition and reduction are described in [5,7]. Reduced spectra are shown in Fig. 2 for 2MASS 0559-14 [4] and Gl 570D [4], along with data for Gl 229B [35] and L dwarfs 2MASS 1507-16 (L5V), 2MASS 0920+35 (L6.5V), and 2MASS 1632+19 (L8V) [19,18]. Spectra are displayed on a log scale and offset in order to highlight features.

A few general trends are readily apparent. Most obvious is the increased spectral slope from 8000–10100 Å in the T dwarfs, a feature which has been attributed to pressure broadened K I at 7665 and 7699 Å [9,25]. A drop in flux shortward of 6600 Å is due to the similarly broadened Na I D lines at 5890 and 5895 Å [37]. Increased H_2O absorption at 9250 Å is seen in the T dwarfs, strongest in Gl 229B and Gl 570D. CrH (8611 Å), FeH (8692 Å), and Na I (8183, 8195 Å) features weaken in the late L dwarfs but are completely absent in the T dwarfs. The behavior of the Cs I (8543, 8943 Å) lines is less straightforward,

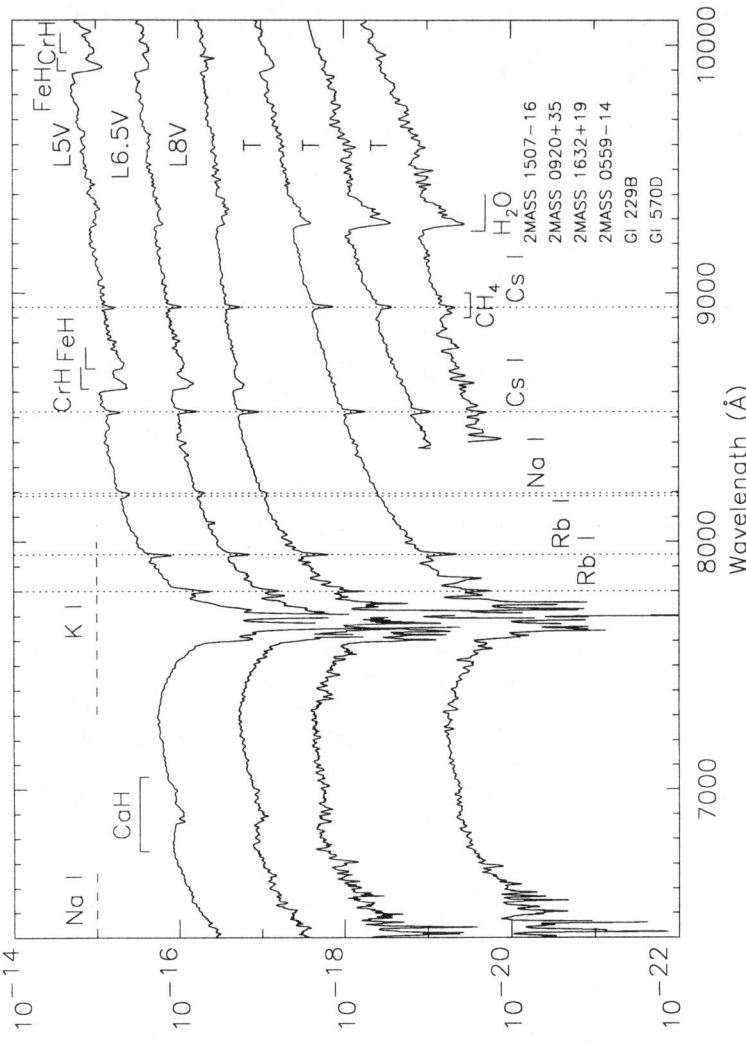

Fig. 2. Red optical spectra of L and T dwarfs. Spectra are displayed on a log scale and offset by a constant factor. Major molecular and atomic features are noted. L dwarf spectra are from [19,18], Gl 229B data from [35].

appearing to strengthen slightly from L5V to 2MASS 0559-14, but weakening in Gl 229B and Gl 570D, with the 8943 Å line becoming blended with CH_4 at 8900 Å. Rb I (7800, 7948 Å) lines may be strengthening from mid-L to 2MASS 0559-14, but lack of signal prevents their detection in the other T dwarfs (only five objects in our sample exhibit flux shortward of 8000 Å). FeH (9896 Å) is known to weaken from mid- to late-L [19] and is barely detected in 2MASS 1632+19; however, 2MASS 0559-14 shows a clear band, while Gl 229B and Gl 570D do not. CaH (6750 Å) may be present in all of the spectra shown here, but is exceedingly weak in the latest L dwarfs and 2MASS 0559-14.

It is clear, then, that there are qualitative differences in the optical spectra of L and T dwarfs. The T dwarfs themselves appear to be distinguishable, based on their spectral slope, the presence of 9896 Å FeH, and the 9250 Å H_2O band strength. To explore these differences quantitatively, we have examined a set of spectral indices, as described in Table 1. Index values are plotted in Fig. 3 for the five brightest T dwarfs in our sample and Gl 229B (open circles), as well as late-M and L dwarf standards from [19] (solid circles). T dwarfs are ordered by their H_2O index. The deepening of the 9250 Å H_2O band in the T dwarfs is readily apparent, as is the increased spectral slope measured by the Colour-d index. Weakening of the 8611 Å CrH and 9896 Å FeH bands from types L to T is similarly seen in their respective indices. The KI-a index, which measures the relative depth of the broadened K I doublet (or alternately its breadth), peaks at around L8V, then decreases through the T dwarfs; this is either due to lower signal-to-noise in the latter objects or the formation of KCl at $T_{eff} \sim 950$ K [26]. Cs I (8543 Å) shows a general strengthening but with significant scatter, reflecting its observed ambiguous behavior.

Table 1. T Dwarf Red Optical Spectral Indices

Index	Flux Ratio	Feature Measured
KI-a	$\frac{7100.0 \to 7300.0}{7600.0 \to 7800.0}$	7665, 7679 Å K I doublet
Cs-a	$\frac{(8496.1 \to 8506.1)+(8536.1 \to 8546.1)}{2\times(8516.1 \to 8526.1)}$	8543 Å Cs I line[a]
CrH-a	$\frac{8580.0 \to 8600.0}{8621.0 \to 8641.0}$	8611 Å CrH band[a]
H_2O	$\frac{9225.0 \to 9250.0}{9275.0 \to 9300.0}$	9250 Å H_2O band
FeH-b	$\frac{9863.0 \to 9883.0}{9908.0 \to 9928.0}$	9896 Å FeH band[a]
Colour-d	$\frac{9675.0 \to 9875.0}{7350.0 \to 7550.0}$	spectral slope[a]

[a] Also used as indices for L dwarfs in [19].

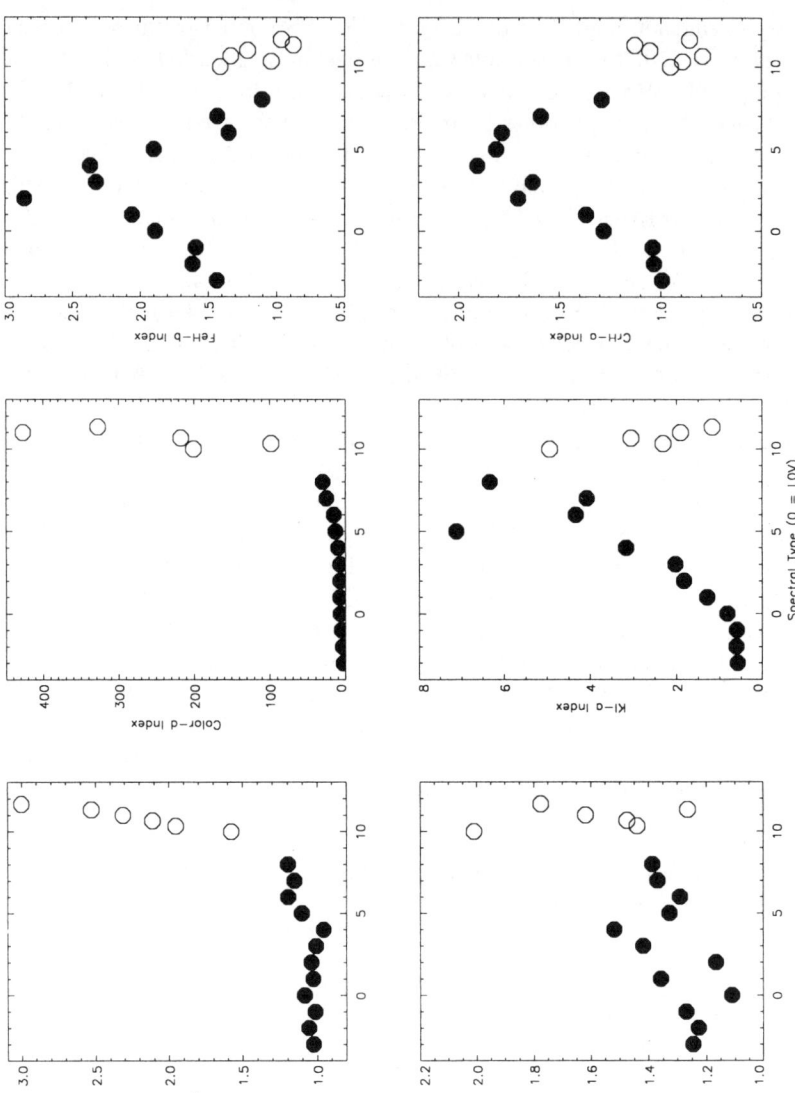

Fig. 3. Red optical spectral indices for late-M, L, and T dwarfs. M and L dwarf values (filled circles) are calculated using data from [19,18]. T dwarfs (open circles) are ordered by their H_2O index. Indices are described in Table 1.

Based on these spectral indices and the quality of the observed spectra, we make three conclusions regarding T dwarf classification in the optical:

- L and T dwarfs (later than 2MASS 0559-14) appear to be distinguishable in the red optical, due to the absence of CrH (8611 Å) and FeH (8692 Å) bands, increased 9250 Å H_2O absorption, and increased spectral slope in the T dwarfs.
- T dwarfs can be separated in the red optical based on the presence or absence of the 9896 Å FeH band, 9250 Å H_2O band strength, and spectral slope. The observed scatter, however, restricts us to distinguish only "early" T dwarfs (e.g., 2MASS 0559-14) from "late" ones (e.g., Gl 570D).
- Despite these successes, the faintness of these objects in the optical ($M_R \sim$ 24.6 for Gl 229B [15]) makes it necessary to consider classification in the near-infrared.

3 Near-Infrared Classification

The spectral energy distribution of T dwarfs peaks at around 1.27 μm, the center of the J-band window, due to the combined effects of temperature and H_2O, CH_4, and collision-induced (CIA) H_2 absorption features. The 1–2.5 μm region also encompasses the defining 2.2 μm CH_4 band. In our program to identify these cool brown dwarfs, we have acquired 1–2.5 μm spectra of 14 T dwarfs at a resolution of R $\sim$ 100 (100Å at 1 μm) using the Near-Infrared Camera (NIRC [28]) mounted on the Keck 10m Telescope. Instrumental setup and data reduction are described in [7]. Figure 4 displays these spectra (thick lines), along with NIRC data for Gl 229B [35], and SDSS T dwarf data from the literature [41,24] (higher resolution thin lines). A NIRC spectrum of 2MASS 0920+35 (L6.5V) is shown for comparison. All spectra are normalized at their J-band peaks.

The spectra in Fig. 4 are ordered by visual inspection, based on the increasing strengths of the 1.15 and 1.6 μm absorption bands. The 1.15 μm feature begins in the late L dwarfs as a blend of H_2O, two K I doublets (1.169 & 1.177, 1.243 & 1.252 μm), Na I (1.138, 1.141 μm), and FeH (1.19, 1.21, 1.24 μm) [29], but is later influenced by CH_4 absorption at 1.1 μm in the T dwarfs. The deepening and broadening of this feature and the adjacent H_2O/CH_4 absorption at 1.4 μm causes the J-band flux to become more peaked and narrow toward the latest objects shown. The 1.15 μm feature also depresses the 1.243 & 1.252 μm K I doublet, unresolved in the NIRC data. There is significant suppression of the H- and K-band peaks between SDSS 1021-03 and 2MASS 2254+31, coinciding with a large near-infrared colour difference between these objects (J-H $\sim$ 0.5 mag redder in SDSS 1021-03). This occurs after CO at 2.3 μm, a prominent feature in late-M and L dwarfs, disappears. Toward later types, the K-band peak shape evolves from two bandheads at 2.2 (CH_4) and 2.3 μm (CO), to a rounded peak with a kink at 2.17 μm, to a symmetric peak at 2.07 μm in 2MASS 0415-09.

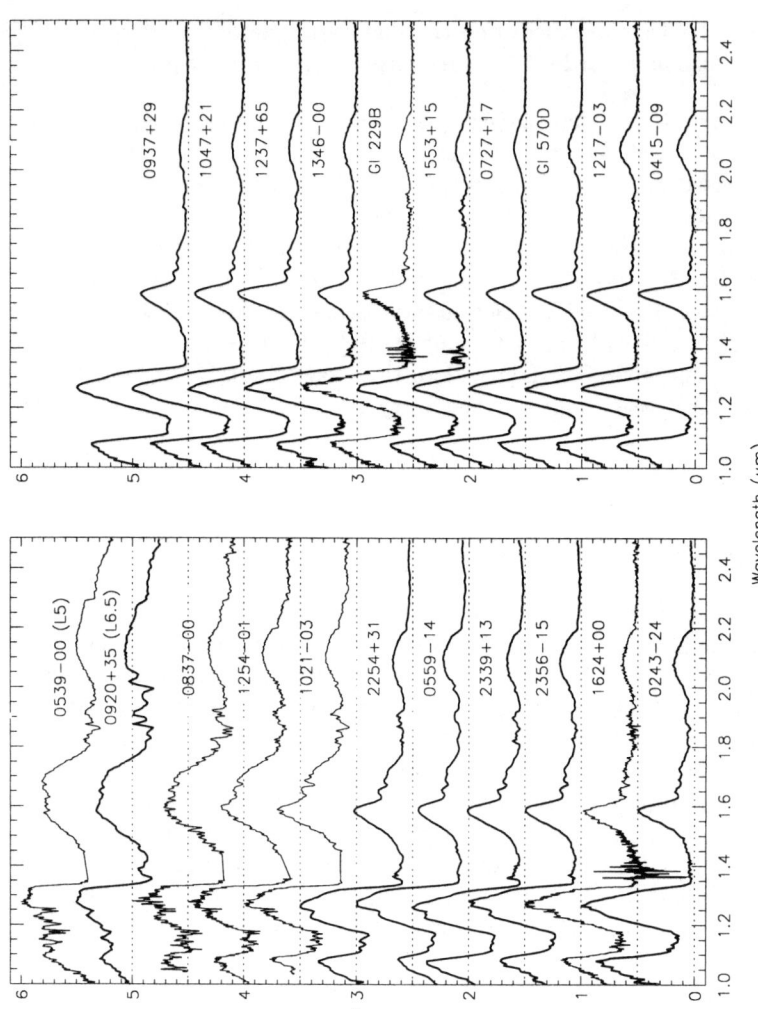

Fig. 4. Near-infrared T dwarf spectra. Data obtained from NIRC observations are shown in thick lines, while (higher resolution) data obtained from the literature [14,41,24] are shown as thinner lines. Spectra are normalized at their J-band peaks and offset by a constant. Order is determined visually by comparison of the depths of the 1.15 μm H_2O/CH_4 and 1.6 μm CH_4 features.

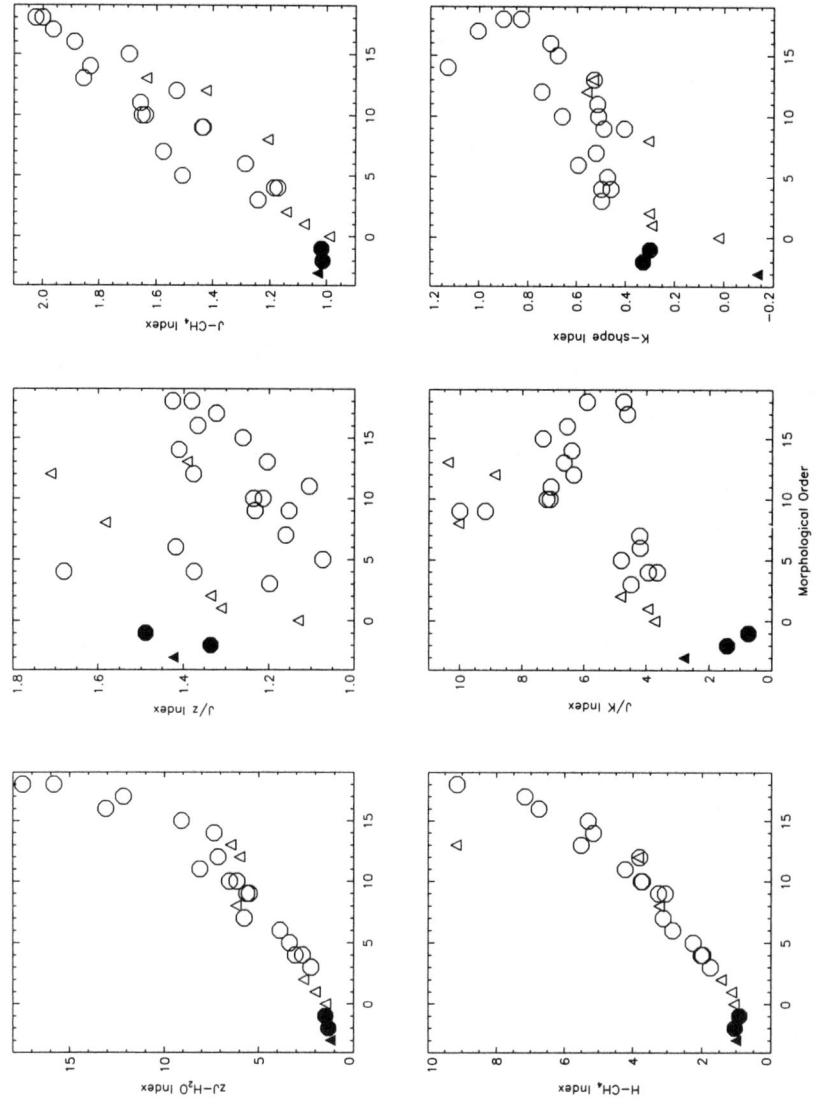

Fig. 5. Near-infrared spectral indices for L (filled symbols) and T dwarfs (open symbols). Measurements of NIRC data are shown as circles, while data obtained from the literature [14,41,24] are plotted as triangles. T dwarfs are ordered as shown in Fig. 4. Indices are described in Table 2.

In order to quantitatively explore these observed spectral variations, we have defined near-infrared spectral indices in analogy to those derived for the red optical spectra. These indices are described in Table 3 and plotted in Fig. 5 for the NIRC data (circles), along with values computed from L and T dwarf spectral data from the literature [14,41,13,24] (triangles). Filled and open symbols denote L and T dwarfs, respectively. T dwarfs are ordered as in Fig. 4, while L dwarfs are ordered by their spectral type from [19]. Aligned symbols represent multiple spectra for the same object, allowing us to estimate the approximate noise in these values, as well as compare spectra of different resolutions. There is an obvious correlation between the depth of the 1.15 μm H_2O/CH_4 band, the 1.3 μm CH_4 band wing, and the 1.6 μm CH_4 band, all of which steadily increase from the late-L dwarfs through 2MASS 0415-09. The K-shape index (measuring the change of the 2.17 μm kink) also increases toward unity (no kink) over this range, although with more scatter. Similarly, the J/K peak ratio appears to generally increase, although there is less agreement between the NIRC data and data from the literature. One object, 2MASS 0937+29 (see below) is particularly discrepant. Finally, the J/z peak ratio shows the least correlation with morphological order, with variations no more than 40% of the mean, suggesting noise as a possible contributor.

Overall, their appears to be a good correlation in the strengths of the CH_4 and H_2O absorption band strengths, and these are likely to be crucial features for spectral typing in the near-infrared. K I lines, unresolved in this data, are also important indicators of temperature (being higher energy lines than the broadened 7665 & 7699 Å K I doublet in the red optical), so that classification spectra in this wavelength regime should be of sufficient resolution to permit measurement of these lines. These data can be easily obtained using existing instrumentation.

Table 2. T Dwarf Near-Infrared Spectral Indices

Index	Flux Ratio	Feature Measured
zJ-H_2O	$\frac{1.25 \to 1.28}{1.13 \to 1.16}$	1.15 μm H_2O/CH_4 band
J/z	$\frac{1.26 \to 1.28}{1.06 \to 1.08}$	J-band/z-band peak ratio
J-CH_4	$\frac{1.26 \to 1.275}{1.295 \to 1.31}$	1.3 μm CH_4 absorption wing
H-CH_4	$\frac{1.57 \to 1.59}{1.63 \to 1.65}$	1.6 μm CH_4 band
J/K	$\frac{1.26 \to 1.28}{2.06 \to 2.09}$	J-band/K-band peak ratio
K-shape	$\frac{(2.10 \to 2.11)-(2.16 \to 2.17)}{(2.17 \to 2.18)-(2.19 \to 2.20)}$	K-band shape (2.17 μm kink)

4 Additional Classification Parameters in the Near-Infrared

The discrepancy seen in the J/K peak ratio for 2MASS 0937+29 is fairly obvious in Fig. 5, and can be seen in Fig. 6a which shows the 1.5–2.5 μm spectra of 2MASS 0937+29 and 2MASS 1047+21, both normalized at their J-band peaks. The similarity of these objects at H-band is in striking contrast to the notable suppression of the K-band peak of 2MASS 0937+29. This object is the bluest T dwarf so far identified, with J-K_s = -0.89±0.24, as compared to the mean J-K_s ~ 0 for all other 2MASS T dwarfs. The K-band peak is shaped by the combined effects of H_2O (1.8–2.1 μm), CH_4 (2.2–2.6 μm), and CIA H_2 (centered around 2.5 μm). The latter feature is strongest at K-band, and, being a collisional process, is highly sensitive to the total photospheric pressure. Zero-metallicity models show that increased gravity (and therefore photospheric temperature) results in bluer H–K colours due to increased H_2 absorption, an effect which becomes more pronounced toward cooler T_{eff} [38].

To examine this effect, we have computed spectral indices for models provided by D. Saumon (priv. comm.). Figure 6b shows the behavior of the J/K peak ratio as compared to 1.6 μm CH_4 band strength for gravities of 300, 1000, and 3000 m s^{-2}. Overplotted are index values measured for the 2MASS T dwarfs (open circles). Gl 570D is separately indicated (solid triangle); this companion object has an estimated T_{eff} = 750±50 K and g = 800–2000 m s^{-2} [4], in consensus with its location in Fig. 6b relative to the models. The large spread in J/K index values versus gravity for the models is obvious, and is greater at lower temperatures. We see that 2MASS 0937+29 and 2MASS 1047+21 (solid circles) lie on opposite sides of the model g = 3000 m s^{-2} line, consistent with the former having a higher surface gravity. This example suggests that comparison of the K-band peaks for a sample of classified T dwarfs may allow segregation by gravity (and therefore mass), in analogy to the dwarf/giant segregation made in the MK luminosity classes. Variations in metallicity may also play an important role in H_2 strength, and the effects of CIA H_2 absorption at other wavelengths (i.e., near peaks at 0.83 and 1.25 μm) needs to be similarly investigated. Alkali lines, particularly the 1.17 and 1.25 μm K I doublets, may also elucidate gravity and/or metallicity effects, and provide a further discriminant for a two-dimensional classification scheme.

5 Summary

We have discussed the prospects for spectral typing T dwarfs in the red optical and near-infrared, based on analysis of spectral data obtained by the authors and from the literature. Basic morphological differences are seen in both regimes, supporting the viability of considering classification schemes for these cool brown dwarfs. Several conclusions may be drawn from the preceding discussion:

- The red optical spectra of T dwarfs is different than that of the L dwarfs, due to differences in spectral slope, 9250 Å H_2O band strength, and the presence

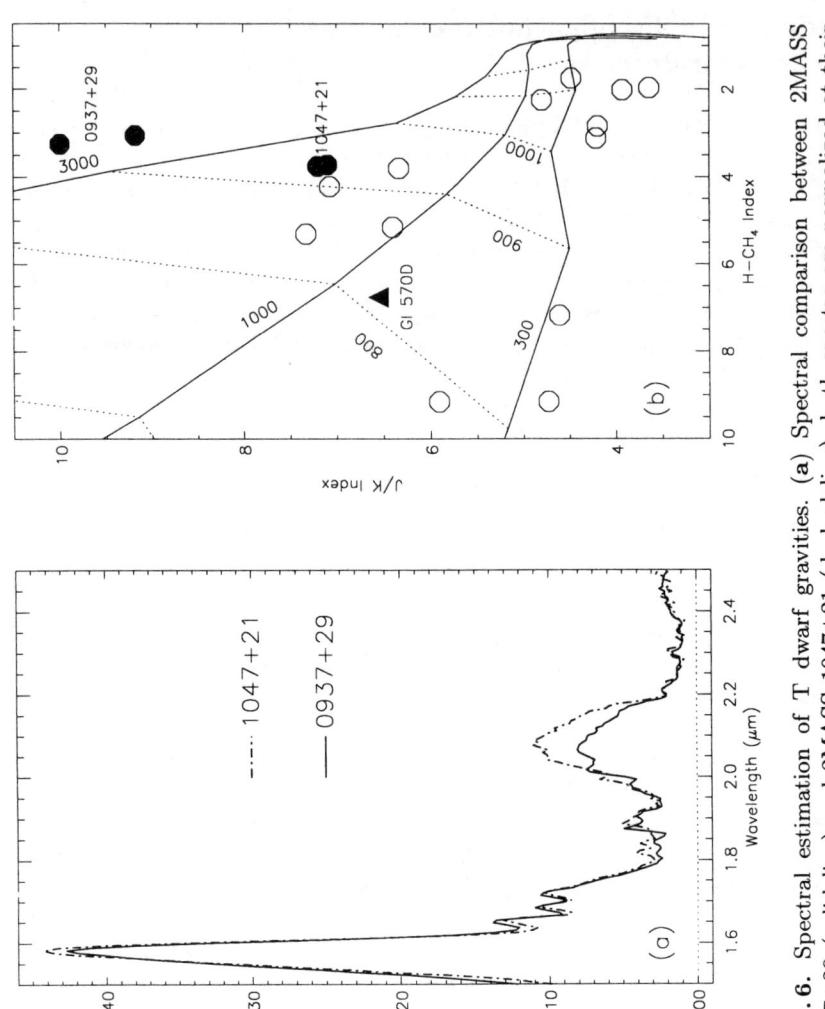

Fig. 6. Spectral estimation of T dwarf gravities. (a) Spectral comparison between 2MASS 0937+29 (solid line) and 2MASS 1047+21 (dashed line); both spectra are normalized at their J-band peaks. (b) H-CH$_4$ index versus J/K peak ratio for 2MASS T dwarfs and T dwarf models provided by D. Saumon (priv. comm.) Lines trace out model values for gravities of 300, 1000, and 3000 m s^{-2}.

or absence of various hydride bands and alkali lines. The faintness of the T dwarfs at these wavelengths, however, favors classification schemes at longer wavelengths.
- A continuous sequence of T dwarfs can be seen in near-infrared spectral data, both at low (R ~ 100) and moderate (R ~ 600) resolutions. This sequence is dominated by the bands of CO (in the earliest T dwarfs), H_2O, CH_4, and CIA H_2, while the 1.25 μm K I doublet, unresolved in the NIRC data, may show a corresponding progression. The clear variation in features and greater facility of obtaining reasonable signal-to-noise spectra favors this wavelength regime for classification.
- Spectral features that are sensitive to specific gravity may allow development of a two-dimensional scheme analogous to the MK luminosity class, but instead based on object mass. The strength of the K-band peak appears to be such a feature, due to pressure-sensitive CIA H_2 absorption. Calibration of this effect and the influence of metallicity remains to be explored.

A.J. Burgasser acknowledges S.K. Leggett, B.R. Oppenheimer, and M.A. Strauss for providing useful spectral data for comparison, and D. Saumon for use of unpublished model spectra. Enlightening discussions over the role of H_2 were had with A. Burrows, M. Marley, and D. Saumon. AJB also acknowledges the contributions of the other members of the 2MASS Rare Objects Team: R. Cutri, C. Dahn, J. Gizis, J. Liebert, B. Nelson, and I.N. Reid. Data presented herein were obtained at the W. M. Keck Observatory which is operated as a scientific partnership among the California Institute of Technology, the University of California, and the National Aeronautics and Space Administration. The Observatory was made possible by generous financial support of the W. M. Keck Foundation. This contribution makes use of data from the Two Micron All Sky Survey, which is a joint project of the University of Massachusetts and the Infrared Processing and Analysis Center and funded by the National Aeronautics and Space Administration and the National Science Foundation.

References

1. P.C. Boeshaar: The spectral classification of M-dwarf stars. PhD Thesis, Ohio State University, Columbus (1976)
2. P.C. Boeshaar, & J.A. Tyson: AJ **90**, 817 (1985)
3. A.J. Burgasser, et al.: ApJ **522**, L65 (1999)
4. A.J. Burgasser, et al.: ApJ **531**, L57 (2000a)
5. A.J. Burgasser, J.D. Kirkpatrick, I.N. Reid, J. Liebert, J.E. Gizis, & M. Brown: AJ **120**, 473 (2000b)
6. A.J. Burgasser, et al.: ApJ **120**, 1100 (2000c)
7. A.J. Burgasser, et al.: ApJ (2001) in prep.
8. A. Burrows, et al.: ApJ **491**, 856 (1997)
9. A. Burrows, M.S. Marley, C.M. Sharp: ApJ **531**, 438 (2000)
10. A.J. Cannon, & E.C. Pickering: Ann. Astron. Obs. Harvard Coll. **28**(II), 131 (1901)
11. J.G. Cuby, P. Saracco, A.F.M Moorwood, S. D'Odorico, C. Lidman, F. Comerón, & J. Spyromilio: A&A **349**, L41 (1999)

12. N. Epchtein, et al.: In *Science with Astronomical Near-Infrared Sky Surveys*. ed. by N. Epchtein, A. Omont, B. Burton B., & P. Persi (Kluwer, Dordrecht 1994) p. 3
13. X. Fan, et al.: AJ **119**, 928 (2000)
14. T.R. Geballe, S.R. Kulkarni, C.E. Woodward, & G.C. Sloan: ApJ **467**, L101 (1996)
15. D.A. Golimowski, C.J. Burrows, S.R. Kulkarni, B.R. Oppenheimer, & R.A. Brukardt: AJ **115**, 2579 (1998)
16. D. Hoffleit: Ann. Astron. Obs. Harvard Coll. **105**, 45 (1937)
17. H.L. Johnson, & W.W. Morgan: ApJ **117**, 313 (1953)
18. J.D. Kirkpatrick, I.N. Reid, J. Liebert, J.E. Gizis, A.J. Burgasser, D.G. Monet, C.C. Dahn, B. Nelson, & R.J. Williams: AJ **120**, 447 (2000)
19. J.D. Kirkpatrick, et al.: ApJ **519**, 802 (1999)
20. J.D. Kirkpatrick, T.J. Henry, & M.J. Irwin: AJ **113**, 1421 (1997)
21. J.D. Kirkpatrick, T.J. Henry, & D.W. McCarthy: ApJ Supp. **77**, 417 (1991)
22. G.P. Kuiper: ApJ **88**, 429 (1938)
23. G.P. Kuiper: ApJ **95**, 201 (1942)
24. S.K. Leggett, et al.: ApJ **536**, L35 (2000)
25. J. Liebert, I.N. Reid, A. Burrows, A.J. Burgasser, J.D. Kirkpatrick, & J.E. Gizis: ApJ **533**, L155 (2000)
26. K. Lodders: ApJ **519**, 793 (1999)
27. E.L. Martín, X. Delfosse, G. Basri, B. Goldman, T. Forveille, & M.R. Zapatero-Osorio: AJ, **118**, 2466 (1999)
28. K. Matthews, & B.T. Soifer: In *Infrared Astronomy with Arrays: The Next Generation*. ed. by I. McLean (Dordrecht, Kluwer 1994) p. 239
29. I.S. McLean, et al.: ApJ, **533**, L45 (2000)
30. W.W. Morgan, P.C. Keenan, & E. Kellman: *An Atlas of Stellar Spectra, with an Outline of Spectral Classification*. (Chicago, Univ. Chicago Press 1943)
31. T. Nakajima, B.R. Oppenheimer, S.R. Kulkarni, D.A. Golimowski, K. Matthews, & S.T. Durrance: Nature **378**, 463 (1995)
32. E.C. Pickering: Ann. Astron. Obs. Harvard Coll. **27**, 1 (1890)
33. J.B. Oke, et al.: PASP **107**, 375 (1995)
34. B.R. Oppenheimer, S.R. Kulkarni, K. Matthews, & T. Nakajima; Science **270**, 1478 (1995)
35. B.R. Oppenheimer, S.R. Kulkarni, K. Matthews, & M.H. van Kerkwijk: ApJ **502**, 93 (1998)
36. R. Rebolo, M.R. Zapatero-Osorio, & E.L. Martín: Nature **377**, 129 (1995)
37. I.N. Reid, et al.: ApJ **521**, 631 (1999)
38. D. Saumon, P. Bergeron, J.I. Lunine, W.B. Hubbard, & A. Burrows: ApJ **424**, 333 (1994)
39. A. Secchi: CR Acad. Sci. Paris **63**, 621 (1866)
40. M.F. Skrutskie, et al.: In *The Impact of Large-Scale Near-IR Sky Surveys*. ed. by F. Garzon (Dordrecht, Kluwer 1997) p. 25
41. M.A. Strauss, et al: ApJ **522**, L61 (1999)
42. Z.I. Tsvetanov, et al: ApJ **531**, L61 (2000)
43. D.G. York, et al: AJ **120**, 1579 (2000)

L-Band Photometry and Spectroscopy of L and T Dwarfs: Exploring Infrared Spectral Typing

D.C. Stephens[1], M.S. Marley[2], and K.S. Noll[3]

[1] New Mexico State University, Las Cruces, NM 88003, USA
[2] Space Sciences Division, NASA/Ames Research Center, Moffett Field, CA 94035, USA
[3] Space Telescope Science Institute, Baltimore, MD 21218, USA

Abstract. The classification of L and T dwarfs requires the identification of easily observed quantities such as colour ratios or strengths of spectral features. Although current classification schemes based on optical spectroscopy for L dwarfs do exist, it is important to classify L and T dwarfs using other wavelengths as well. Observations of L and T dwarfs at infrared wavelengths are sensitive to physical processes controlling the atmospheres and temperatures of brown dwarfs which provide for the possibility of L and T dwarf classification using infrared observations. Here we present the first results of our observing program using the Keck telescope to obtain infrared photometry and spectroscopy of L and T dwarfs.

1 Introduction

Long elusive to direct detection, brown dwarfs were nothing more than a theoretical construct until the discovery of the first indisputable brown dwarf, Gliese 229B in 1995 [1], whose sub-stellar nature was confirmed by strong methane features in the near-infrared. Since this momentous discovery, the 2-Micron All-Sky Survey (2MASS), the Deep Near-Infrared Southern Sky Survey (DENIS), and the Sloan Digital Sky Survey (SDSS) have combined together to successfully discover over 150 low mass objects with spectral types later than "M". These discoveries have established the need for two new spectral classes: "L" and "T" [2]. The L dwarfs are characterized by the progressive disappearance of TiO and VO bands as these molecules precipitate out of the atmosphere into dust grains, grain opacity, the presence of silicate clouds, and effective temperatures below ~2200 K. The much cooler T dwarfs are characterized by the appearance of methane absorption at 1.6 and 2.2 μm, relatively clear atmospheres, and temperatures below ~1200 K.

The process of characterizing, or classifying brown dwarfs, both L and T, is ultimately a process of understanding important physical parameters and their expression in easily observed quantities such as colour ratios or strengths of spectral features. Although uncertainties remain, particularly for the latest L dwarfs, the classification of L dwarfs is well underway with two competing schemes proposing 6-8 spectral subclasses L dwarfs [2,3]. Both of these schemes use spectral features observed in the far optical (~0.6 to 1 μm) for classification purposes. For T dwarfs, classification has yet to begin, both because of the heretofore small number of objects and the difficulty of applying the same optical wavelength criteria used for L dwarfs. Although current classification schemes for L dwarfs do exist, it's important that we image and classify L and T dwarfs in other wavelength regimes, such as the near-infrared, to understand the physical processes which underlie the spectral features and colour ratios that are observed.

The use of near-infrared photometry and spectroscopy to classify L and T dwarfs may provide a powerful alternative and complement to the current optical classification of L dwarfs. Although optical classification is commonly used, it is impractical to classify the faintest objects and it provides little information about the most abundant atmospheric species such as water (H_2O), carbon monoxide (CO), and methane (CH_4). Since brown dwarfs are brighter in the near-infrared where these abundant molecules have strong absorption bands, a desirable alternative is to observe and classify L and T dwarfs based on their near-infrared photometric and spectroscopic properties. To determine the utility of near-infrared spectral classification, we have obtained K- and L-band photometry for 18 L dwarfs ranging in spectral type from L0 to L8, 3 early T dwarfs, and 2 late T dwarfs using the Keck I telescope. In addition we obtained grism spectroscopy from 2 to 4 μm for 4 late L dwarfs and 1 early T dwarf. Here we present the results of our observations which are part of an ongoing program to obtain photometry and spectroscopy of L and T dwarfs in the near-infrared and also to characterize and understand the atmospheres of these objects using near-infrared observations.

2 Filter Selection

The K, Ks, Ls and L' filters were selected due to their ability to identify the first appearance of methane at 3.3 μm. Understanding the onset of methane absorption in brown dwarf atmospheres is crucial as we try to determine the effective temperature scale for L and T dwarfs. The onset of methane will first be detected in the Ls band at 3.3 μm, where the ν_3 fundamental band is located. This band is two orders of magnitude stronger than the $2\nu_3$ overtone band (1.66 μm) and combination bands found in the H and K windows (Fig. 1). Based on theoretical models and estimated temperatures for L dwarfs, methane is predicted to be visible in the very latest L dwarfs. Indeed Noll et al. [4] reported the appearance of methane in two late L dwarfs at 3.3 μm. By obtaining photometry and spectroscopy near 3.3 μm for several different L sub-classes, we can determine at what sub-class methane first appears. This leads to an improved understanding of the physical processes which are at work in brown dwarf atmospheres since the appearance and abundance of methane provides clues to the atmospheric temperature, cloud micro-physical processes and non-equilibrium chemistry.

L-band photometry aids our understanding of the existence and properties of clouds. Theoretical models of the observed spectra of L dwarfs show clear signs of grain opacity [5]; whereas T dwarfs have very clear atmospheres with few con-

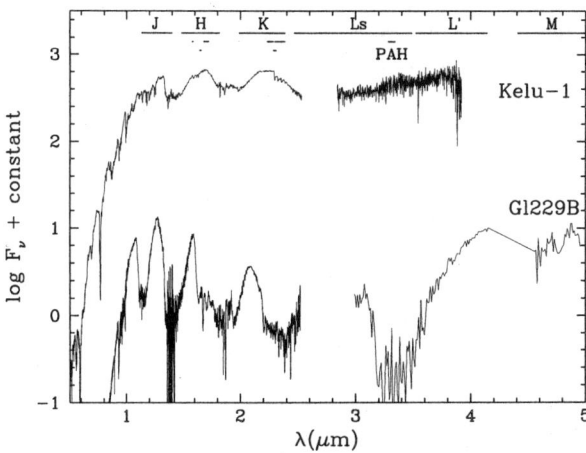

Fig. 1. Spectra of Kelu I (~1900 K) and Gliese 229B (~950 K) with the near-infrared broad and narrow band NIRC filters. Note that the strong ν_3 fundamental band of methane (3.3 μm) is two orders of magnitude stronger than the $2\nu_3$ overtone band (1.66 μm) and combination bands found in the H and K filters. Also note that the Ls and PAH filters directly probe the appearance of methane at 3.3 μm. (Kelu I spectra courtesy of S. Leggett and K. Noll; Gliese 229B spectra courtesy of S. Leggett and B. Oppenheimer)

densates. This difference exists because the silicate clouds found in the hotter L dwarfs form below optical depth unity in the cooler T dwarfs. Therefore at some point in the progression from the late L dwarfs to the early T dwarfs a transition occurs from a cloudy to a clear atmosphere. Understanding the existence and properties of clouds is crucial to determine the effective temperature of L and T dwarfs since the presence of a cloud in the observable atmosphere of a brown dwarf can warm the atmosphere through a greenhouse effect thereby changing the equilibrium abundance of methane. Thus the effects of clouds and methane formation must be disentangled from one another to estimate accurately the effective temperature of brown dwarfs.

Observations of L and T dwarfs using L-band photometry is one way to disentangle the effects of clouds from methane. From studies of the Jovian atmospheres, we know that the radiative effect of cloud particles depends strongly on wavelength. For example, grains with a radius ~1 μm and smaller will provide very little opacity at longer wavelengths. Silicate clouds with particles near 1 μm in size and smaller, would effect the visible spectrum and the J and H bands, while the K and L bands would be less-strongly influenced. Preliminary models provided by Marley support this idea showing that the $(J\text{-}K)$ colour is strongly perturbed by the presence of clouds and that a cloudy atmosphere will differ from a clear atmosphere of the same temperature by a colour index of 2. In contrast, the $(K\text{-}L')$ colour index will differ from a cloudy to a clear atmosphere by 0.2 at most (Fig. 2a & 2b). Thus observations in L' may be used to determine the effective temperature of objects because this filter is relatively unaffected by clouds, while observations using J, H, and K probe the micro-physical properties of clouds.

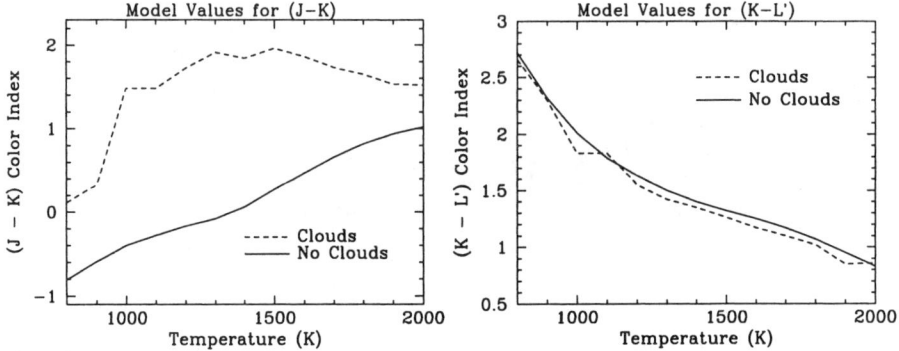

Fig. 2. Clear and cloudy preliminary atmosphere models. In this example the presence of clouds substantially affects the (J-K) colour. However, clouds do not affect the (K-L') colours, thus (K-L') colours can be used to constrain the effective temperature of L and T dwarfs. Also notice the rapid change in (K-L') colour from 800 K to 1200 K where T dwarfs exist. This colour shows promise to classify T dwarfs

3 Observations

The data were obtained on April 13th, 14th, and 15th, UT 2000 using the Near Infrared Camera (NIRC) [6] on the 10m W.M. Keck I Telescope in Mauna Kea, Hawaii. Table 1 contains the list of objects which were observed, their coordinates, spectral classifications, magnitudes and colours. Photometry was acquired using the K, Ks, Ls, L' and PAH filters (fig. 1). The greatest challenge faced in acquiring the photometry was overcoming the high sky background in L' and Ls which limited integration times to 0.012 and 0.03 seconds per exposure respectively. To acquire these short integration times, the pat4xa timing pattern was used and numerous coadds were taken to achieve a high signal-to-noise ratio [7]. Since the sky background is highly variable beyond 3 μm, standard chopping and nodding procedures were employed to minimize errors due to instabilities in the sky background. Sky frames were used to correct for dark current, to divide out instrumental effects, and to subtract off the sky background. Magnitudes were calculated using standard IRAF photometric routines, and several standard stars were used to correct for atmospheric effects.

Spectroscopy was also obtained for four late L dwarfs (L5, L6.5, L7.5, L8) and 1 early T dwarf using the *gr60* grism with the *KL* blocker on the NIRC camera. This provided 2 to 4 μm spectra. The targets were placed in the lower half of a 2.5 pixel slit and nodded up and down along the slit to obtain spectra at

Table 1. Colours and Magnitudes for Observed Brown Dwarfs

Object Name	Spectral Type	K	L'	$K-L'$	$K-L_s$
TVLM 51346	M8.5	10.73 ± 0.003	9.92 ± 0.008	0.81 ± 0.008	0.12 ± 0.005
DENIS J1159+0057	L0	12.80 ± 0.008	11.92 ± 0.033	0.88 ± 0.034	0.18 ± 0.033
DENIS J1441−0945	L1	12.65 ± 0.007	11.77 ± 0.075	0.89 ± 0.075	−1.10 ± 0.089
2MASSW J1035+2507	L1	13.34 ± 0.005	12.27 ± 0.072	1.07 ± 0.072	−0.86 ± 0.184
2MASSW J1411+3936	L1.5	13.26 ± 0.003	12.12 ± 0.068	1.13 ± 0.068	0.19 ± 0.046
2MASSW J0928−1603	L2	13.63 ± 0.004	13.09 ± 0.208	0.53 ± 0.208	0.24 ± 0.068
2MASSW J1338+4140	L2.5	12.75 ± 0.004	11.84 ± 0.059	0.91 ± 0.059	0.22 ± 0.040
2MASSW J1615+3559	L3	12.93 ± 0.003	11.90 ± 0.050	1.03 ± 0.050	0.28 ± 0.036
2MASSW J1246+4037	L4	13.24 ± 0.004	12.04 ± 0.036	1.21 ± 0.036	0.26 ± 0.058
2MASSW J1155+2307	L4	14.10 ± 0.004	12.84 ± 0.114	1.26 ± 0.114	0.43 ± 0.047
2MASSW J1112+3548	L4.5	12.70 ± 0.006	11.51 ± 0.032	1.19 ± 0.033	0.36 ± 0.028
2MASSW J1328+2114	L5	14.23 ± 0.005	13.19 ± 0.074	1.04 ± 0.074	0.41 ± 0.054
2MASSW J1553+2109	L5.5	14.72 ± 0.006	13.19 ± 0.073	1.52 ± 0.073	0.51 ± 0.071
2MASSI J0756+1254	L6	14.91 ± 0.013	13.57 ± 0.067	1.34 ± 0.069	0.30 ± 0.072
2MASSW J0829+2655	L6.5	14.90 ± 0.016	13.47 ± 0.134	1.43 ± 0.134	0.69 ± 0.055
2MASSI J1526+2043	L7	13.92 ± 0.005	12.60 ± 0.094	1.32 ± 0.094	0.41 ± 0.046
2MASSI J0825+2115	L7.5	13.01 ± 0.004	11.46 ± 0.015	1.55 ± 0.015	0.45 ± 0.033
2MASSW J1632+1904	L8	14.02 ± 0.005	12.58 ± 0.094	1.44 ± 0.094	0.52 ± 0.039
SDSS 0837−0000	T	15.96 ± 0.025	14.45 ± 0.126	1.51 ± 0.129	0.08 ± 0.095
SDSS 1254−0122	T	13.95 ± 0.004	12.15 ± 0.049	1.80 ± 0.049	0.37 ± 0.062
SDSS 1021−0304	T	15.48 ± 0.005	13.55 ± 0.112	1.94 ± 0.112	0.18 ± 0.113
2MASSW J1346−0031	T	16.03 ± 0.006	13.87 ± 0.052	2.15 ± 0.052	0.67 ± 0.158
SDSS 1624+0029	T	15.74 ± 0.012	13.19 ± 0.076	2.54 ± 0.076	0.26 ± 0.221

two different locations in an ABBA pattern. All of the data were reduced using standard IRAF routines. The A and B frames were first subtracted from one another and then the spectra were extracted from each frame. The nearby sky background was also extracted from the frames and subtracted from the spectra. A standard star spectra was used to divide out instrumental and atmospheric effects. The wavelength correction was determined by comparing standard star spectra with previously calibrated spectra and then applying the correction to all of the objects. Finally, the spectra were coadded to produce one final spectrum for each object (see Fig. 3). The most noticeable feature in these spectra is a strong absorption band at 3.5 μm due to the resin used to bond the grism which is not removed by standard calibration procedures. We are carefully re-reducing these spectra in an attempt to understand and remove this feature. The T dwarf spectra does show a very strong methane absorption band at 3.3 μm, but we won't be able to confirm the possible appearance of methane in the late L dwarfs until the resin feature is correctly accounted for and removed. Further analysis of these spectra will be released at some future date when a more thorough reduction is complete.

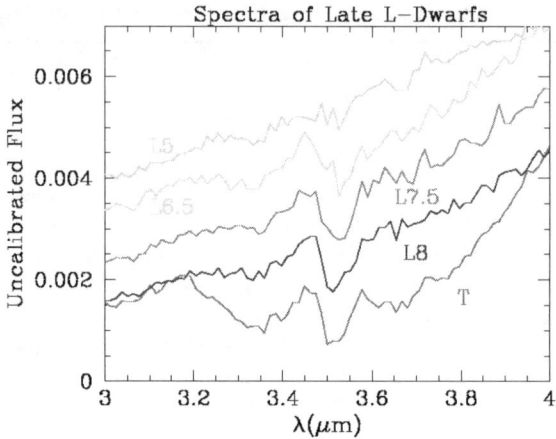

Fig. 3. *L*-band spectra of four late L dwarfs and 1 early T dwarf. Note that the strong absorption feature at 3.5 μm is an instrumental effect due to the resin that was used to create the grism and is not a feature of the brown dwarfs. Note also that the T dwarf has been plotted on top of the L8 dwarf to emphasize the strong methane absorption feature at 3.3 μm in the T dwarf.

4 L-Band Colours and Results

The resulting (K-Ls) and (K-L') colours as a function of spectral type for our L and T dwarfs are seen in Figs. 4 and 5. We find that in general the (K-Ls) colour index gradually increases through the L dwarfs with later spectral type and does not turn over and approach the smaller colour index of the T dwarfs. Since the Ls band directly probes the appearance of methane at 3.3 μm, this result suggests that there is probably very little methane absorption in the late L dwarfs; whereas, the methane band at 3.3 μm in the early T dwarfs appears to be very strong. This result is supported by the spectra in Fig. 3 where we see very little, if any absorption at 3.3 μm in the late L dwarfs, and then a large absorption feature in the early T dwarf. This sudden appearance of methane raises interesting questions as to what is occurring at the L to T transition. Such a sudden appearance of methane in the T dwarfs could imply special atmospheric conditions, such as the sudden disappearance of clouds [4], or an effective temperature range in which brown dwarfs can't exist [8]. The possibility also exists that there may be missing transition objects which bridge the L/T gap and have yet to be found. Whatever the reason, more L-band observations of the latest L dwarfs and the earliest T dwarfs are needed to understand the sudden appearance of methane in T dwarfs at 3.3 μm.

Another trend in infrared colour is seen in Fig. 5 where the (K-L') colour gradually increases with later spectral type from the L dwarfs into the T dwarfs. A linear line has been fit to the L-dwarf spectral sequence using a least-squares fit to emphasize the increase in colour. The observed scatter seen around this

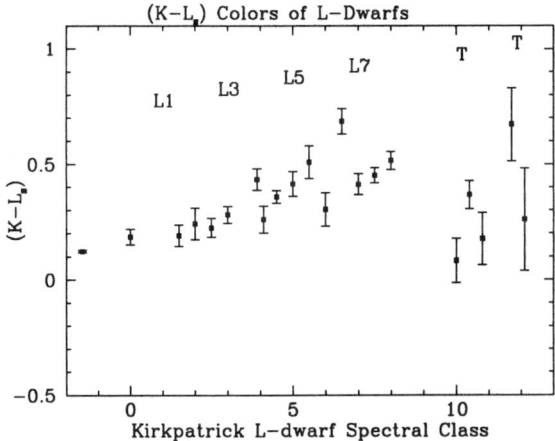

Fig. 4. (K-Ls) colours for the observed L and T dwarfs. We observe a jump in methane abundance from the late L to the the early T dwarfs which produces interesting questions as to why methane suddenly appears in the early T dwarfs, and yet is hardly observable in the latest L dwarfs. The (K-Ls) colours of T dwarfs will be affected by the appearance of methane in both the K and Ls bands, and the difference in colours for the T dwarfs is dependent upon the ratio of methane absorption between the two bands

line may be due to variations in the atmospheres of brown dwarfs. Perhaps differences in cloud structures, the abundance of clouds, or changes in metallicity are contributing to the overall magnitude measured in the K band, and hence the (K-L') colour. For example, increased hydrogen opacity may suppress the H and K flux in low metallicity models [9].

The three T dwarfs seen in fig. 5 with the smallest (K-L') colours are the three L/T transition objects announced by Sandy Leggett et al. [10]. Their location in Fig. 5 confirms that there is less methane absorption in these objects then in the later T dwarfs. The observed increase in (K-L') for the T dwarfs is due to the gradual growth of the methane absorption band at 2.2 μm. This large variation in colour for the T dwarfs shows promise for assigning quickly spectral sub-classification to T dwarfs and assigning effective temperatures to these objects. Particularly those which are too faint or distant to be classified using near-infrared spectroscopy. This colour will also be of use for interpreting extrasolar giant planets (EGPs) since early EGP detections (by chronographic imaging for example) will most likely provide only broad band colours or at very best low-resolution spectra.

5 Comparison of Results with Models

Figure 6 is a plot of the (J-K) versus (K-L') colours for the objects in Table I. The dotted line represents the colours predicted from preliminary cloudy atmo-

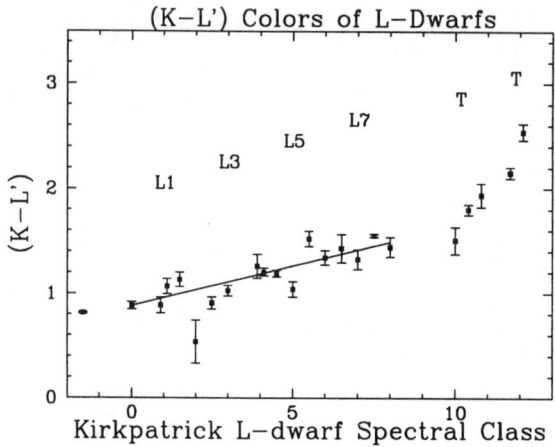

Fig. 5. $(K-L')$ colour for the observed L and T dwarfs. The line indicates a least squares fit that has been applied to the L-dwarf data. Notice that the general trend is for $(K-L')$ to increase throughout the L- and T-dwarf sequences. T dwarfs are particularly bright in L', and $(K-L')$ changes so quickly through the T dwarfs that we believe this colour will be an excellent diagnostic of effective temperature and spectral class

sphere models ranging in temperature from 2000 K to 700 K, and the solid line represents colours predicted from clear models at the same temperatures. It's clear from these results that the cloudy models best reproduce the observed L-dwarf colours, whereas the T-dwarf colours are not well-fit by either a cloudy or a clear atmosphere model. These results support the idea that clouds are prevalent in the atmospheres of L dwarfs and perhaps still play a role in affecting T-dwarf colours.

Another result may be drawn by comparing the observed $(K-L')$ colours with the effective temperatures derived from theoretical models [11]. To produce Fig. 7 we used the least squares fit derived for the L dwarfs in Fig. 5 to determine an approximate $(K-L')$ colour for each L sub-class. Then using those colours, we determined an effective temperature for each L sub-class by locating the predicted temperature at which that colour would appear based on the theoretical models. Then each of the observed data points were assigned to the effective temperature which corresponded to it's spectral type. Finally the observed data points for the L dwarfs were plotted on top of the theoretical models at the point where their $(K-L')$ colours intercepted the estimated effective temperature determined from the $(K-L')$ least squares fit.

The T dwarfs were placed on the plot where their $(K-L')$ colours intercepted the cloudy and clear theoretical models. The resulting T-dwarf sequence is the same as that suggested by others for these T dwarfs [10,12] who based their assumptions on the amount of methane absorption seen in the H and K bands. Once the T-dwarf spectral sequence is defined, we believe that the $(K-L')$ colour will be valuable for assigning T sub-class and effective temperatures to these

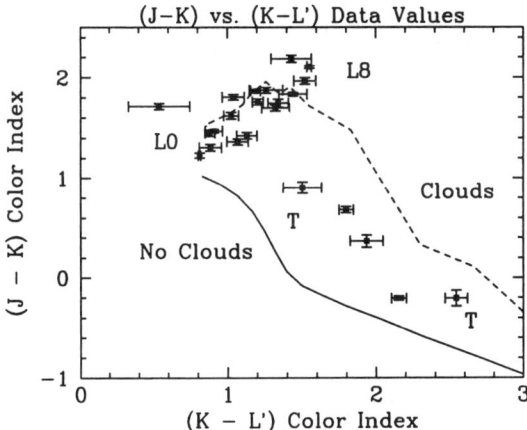

Fig. 6. A plot of $(J\text{-}K)$ vs $(K\text{-}L')$ colours for the observed L and T dwarfs with theoretical models for both cloudy and clear atmospheres plotted on top for comparison. Note that the cloudy models do a relatively good job of fitting the L dwarfs, but that the T dwarfs are not fit well by either a cloudy or a clear atmosphere model

objects. For example, these results support an L to T transition near an effective temperature of ~1200 K. Another interesting result seen in this plot is that all of the latest L-dwarfs from L5.5 to L8 have very similar $(K\text{-}L')$ colours. This result in turn suggests that these objects have very similar effective temperatures which supports the Martin et al. [3] spectral classification claim that the latest L-dwarfs vary in temperature by at most 200 K. Future observations of other late L dwarfs will determine the extent to which this trend holds true.

6 Conclusions

Observations of L and T dwarfs at 3 μm and longer wavelengths are expanding our knowledge of the important physical processes which determine the atmospheric structure and temperature of brown dwarfs. Although this is a difficult spectral region in which to work, there are advantages to be gained in the understanding and classifying of L and T dwarfs. For example, the onset of methane absorption will first appear at 3.3 μm. Through L-band photometry and spectroscopy we can constrain the spectral type at which methane first appears, and since methane probes through equilibrium chemistry and atmospheric temperature, it provides information on greenhouse heating by dust grains, cloud micro-physical processes, and non-equilibrium chemistry. This work shows that if methane does exist in the latest L dwarfs, it is very weak at 3.3 μm; whereas the earliest known T dwarfs show very strong methane features. The relative lack of methane absorption in late L dwarfs may provide information on atmospheric physics and further L-band observations of late L dwarfs will be needed to provide reasonable theoretical explanations, and to confirm these results.

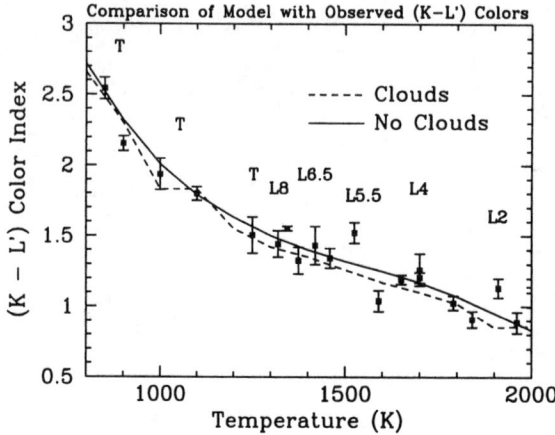

Fig. 7. A plot of (K-L') vs temperature for cloudy and clear atmosphere models with the (K-L') colours of L and T dwarfs plotted for comparison. From this fig. its clear that the (K-L') colour will be a strong diagnostic in determining the effective temperature and spectral classification of T dwarfs. Note also that the latest L dwarfs all have very similar (K-L') colours, and possibly have similar effective temperatures as well.

Observations beyond 3 μm are also less affected by clouds and may provide information about effective temperature and micro-physical cloud properties when combined with observations obtained from the visible and J, H and K bands. Variations in (K-L') colours may reveal information about differences in the atmospheric structure and composition of brown dwarfs. Information about metallicity, grain particle size, and the abundance of clouds may reveal themselves when photometric colours and spectroscopic observations are combined. Photometric colours like (K-L') can also probe objects which are too faint for optical spectroscopy such as T dwarfs and determine their effective temperature and spectral class.

To conclude, the classification and real understanding of L- and T-dwarf atmospheres will occur only after these objects have been observed in several different wavelength regions using both photometry and spectroscopy. Only then can we begin to understand the complex physical processes which control the colour ratios and spectral lines we observe. The L-band observations presented here are just one piece to the puzzle, which when combined with other pieces will provide clues about the physics which controls the temperature and atmospheric structure of brown dwarfs. Further observations and theoretical modelling will clarify our understanding of low mass objects and bring us one step closer to a more complete understanding of the physical nature and characteristics of L and T dwarfs.

Acknowledgments

The data presented herein were obtained at the W.M. Keck Observatory, which is operated as a scientific partnership among the California Institute of Technology, the University of California, and the National Aeronautics and Space Administration. The Observatory was made possible by the generous financial support of the W.M. Keck Foundation.

References

1. T. Nakajima, B.R. Oppenheimer, S.R. Kulkarni, D.A. Golimowski, K. Matthews, S.T. Durrance: Nature. **378**, 463 (1995)
2. J.D. Kirkpatrick, I.N. Reid, J. Liebert, R.M. Cutri, B. Nelson, C.A. Beichman, C.C. Dahn, D.G. Monet, J.E. Gizis, M.F. Skrutskie: ApJ. **519**, 802 (1999)
3. E.L. Martín, X. Delfosse, G. Basri, B. Goldman, T. Forveille, M.R. Zapatero Osorio: AJ. **118**, 2466 (1999)
4. K.S. Noll, T.R. Geballe, S.K. Leggett, M.S. Marley: ApJ **541**, L75 (2000)
5. J.D. Kirkpatrick, F. Allard, T. Bida, B. Zuckerman, E.E. Becklin, G. Chabrier, I. Baraffe: ApJ. **519**, 834 (1999)
6. K. Matthews, B.T. Soifer: In: *Infrared Astronomy with Arrays: The next Generation*, ed. I. McLean (Dordrecht: Kluwer 1994) p. 239
7. W. Harrison, R.W. Goodrich: The NIRC Users Manual. http://www2.keck.hawaii.edu:3636/realpublic/inst/nirc/manual/Manual.html (1999)
8. G. Chabrier, I. Baraffe, F. Allard, P. Hauschildt: ApJ. **542**, 464 (2000)
9. D.C. Stephens, M.S. Marley: 'Narrow Band Near Infrared Photometry of Brown Dwarfs'. In: *From Giant Planets To Cool Stars, ASP Conference Series Volume 212*, ed. by C.A. Griffith, M.S. Marley (PASP, 2000) pp. 187-193
10. S.K. Leggett, T.R. Geballe, X. Fan, D.P. Schneider, J.E. Gunn, R.H. Lupton, G.R. Knapp, M.A. Strauss, A. McDaniel, D.A. Golimowski, T.J. Henry, E. Peng, Z.I. Tsvetanov, A. Uomoto, W. Zheng, G.J. Hill, L.W. Ramsey, S.F. Anderson, J.A. Annis, N.A. Bahcall, J. Brinkmann, B. Chen, I. Csabai, M. Fukugita, G.S. Hennessy, R.B. Hindsley, Z. Ivezić, D.Q. Lamb, J.A. Munn, J.R. Pier, D.J. Schlegel, J.A. Smith, C. Stoughton, A.R. Thakar, D.G. York: ApJ. **536**, L35 (2000)
11. M.S. Marley: 'The role of condensates in the L- and T-dwarf Atmospheres'. In: *From Giant Planets To Cool Stars, ASP Conference Series Volume 212*, ed. by C.A. Griffith, M.S. Marley (PASP, 2000) pp. 152-162
12. Z.I. Tsvetanov, D.A. Golimowski, W. Zehng, T.R. Geballe, S.K. Leggett, H.C. Ford, A.F. Davidson, A. Uomoto, X. Fan, G.R. Knapp, M.A. Strauss, J. Brinkmann, D.Q. Lamb, H.J. Newberg, R. Rechenmacher, D.P. Schneider, D.G. York, R.H. Lupton, J.R. Pier, J. Annis, I. Csabai, R.B. Hindsley, Z. Ivesic, J.A. Munn, A.R. Thakar, P. Waddell: ApJ. **531**, L61 (2000)

Index

accretion, 163
activity, 6, 54, 71, 73, 109, 163
adaptive optics, 59
age estimation, 71, 77, 130, 156, 164
alkali metals, 5, 11, 23, 26, 29, 35, 179

binary fraction, 112, 115
binary systems, 53, 56, 111, 125

companions, 53, 56, 111, 125
coronae, 80

deuterium test, 46, 164
dust, 5, 6, 10, 12, 13, 22, 34, 54, 108, 137

GD 165B, 9, 19, 54, 115
GL 229B, 5, 9, 10, 19, 27, 53, 56, 85, 115, 173, 185
gravity, 137, 159, 179

Hertzsprung-Russell diagram, 3, 121, 135, 163

infrared colours, 14, 53, 83, 86, 137, 139, 184, 188
infrared spectroscopy, 83, 175, 186

kinematics, 77, 79

L/T transition, 5, 9, 86, 90
light curve, 93
lithium test, 45, 53, 77, 130, 154, 164

luminosity function, 121

magnetic spots, 108
mass estimation, 63, 71, 77, 130, 156
methane, 18, 19, 83, 89, 175, 179, 184, 185

open clusters, 53, 78, 154
optical colours, 28, 55, 86, 136

parallax, 3
planets, 55, 57, 65, 156
power spectrum, 95
proper motions, 119

Q-index, 155

rotation, 6, 71, 104

spectral classification, 3, 22, 90, 135, 136, 139, 154, 170, 183
star forming regions, 154
surface features, 104
synthetic colours, 14, 28, 189
synthetic spectra, 17, 34, 35, 137, 157

T Tauri stars, 163

variability, 6, 54, 92

X-rays, 80

**You are one click away
from a world of physics information!**

Come and visit Springer's

Physics Online Library

Books

- Search the Springer website catalogue
- Subscribe to our free alerting service for new books
- Look through the book series profiles

You want to order?   Email to: orders@springer.de

Journals

- Get abstracts, ToC´s free of charge to everyone
- Use our powerful search engine LINK Search
- Subscribe to our free alerting service LINK *Alert*
- Read full-text articles (available only to subscribers of the paper version of a journal)

You want to subscribe?   Email to: subscriptions@springer.de

Electronic Media

- Get more information on our software and CD-ROMs

You have a question on
an electronic product?   Email to: helpdesk-em@springer.de

● Bookmark now:

http://www.springer.de/phys/

Springer · Customer Service
Haberstr. 7 · 69126 Heidelberg, Germany
Tel: +49 (0) 6221 - 345 - 217/8
Fax: +49 (0) 6221 - 345 - 229 · e-mail: orders@springer.de
d&p · 6437.MNT/SFb

Springer

Druck: Strauss Offsetdruck, Mörlenbach
Verarbeitung: Schäffer, Grünstadt